Dynamical Models in Biology

Dynamical Models in Biology

Miklós Farkas

School of Mathematics
Budapest University of Technology
Budapest, Hungary

ACADEMIC PRESS

A Harcourt Science and Technology Company

San Diego San Francisco New York
Boston London Sydney Tokyo

This book is printed on acid-free paper. ♾

ACADEMIC PRESS
A Harcourt Science and Technology Company
525 B Street, Suite 1900, San Diego, CA 92101-4495, USA
http://www.academicpress.com

Academic Press
Harcourt Place, 32 Jamestown Road, London, NW1 7BY, UK
http://www.academicpress.com

Library of Congress Catalog Number: 00-111107

International Standard Book Number: 0-12-249103-3

Printed in the United States of America
01 02 03 04 05 ML 9 8 7 6 5 4 3 2 1

To Kati,
my companion also for the coming 23

Preface

"But you, as a man of science, will have seen
A host of curious things—recall that worm
Which lives only inside a cat or kestrel,
But nonetheless must spend the earliest phase
Of its life-cycle inside a common mouse.
No particular mouse is singled out
To feel the cat's claw, or the kestrel's talon,
One that is careful could avoid them both
And die at home, at a venerable age.
But there's an iron law which so requires,
That there be mice enough to go around
That even the defenceless worm survives
And prospers after many thousand years."

Madách: The Tragedy of Man
G. Szirtes' translation

Twelve years ago as a visiting professor I taught a course on Mathematical Population Dynamics at the Universidad Central de Venezuela in Caracas. For the past 10 years I have been teaching a broader spectrum Biomathematics graduate course at the Budapest University of Technology. This book is a result of these two courses. It is aimed at mathematicians interested in applications as well as biologists, medical doctors, and agricultural engineers who have a somewhat higher than average mathematical background. I do not want to teach biology here; I try to present and treat those *mathematical methods* that are used to describe dynamical phenomena in biology. As much biological explanation of the problems is given as seems to be absolutely necessary for a nonbiologist to understand the situation, and the mathematical formulae are explained intuitively to aid both nonmathematicians and beginning ones. At the same time, I do not want to avoid the really difficult topics. The main mathematical tools applied here are dynamical systems, ordinary and partial differential equations, and bifurcation theory. Rigorous background material is often found in the Appendices. It is assumed that the reader has a knowledge of advanced calculus and linear algebra. Everything else needed is contained in a concise form in the Appendices. I attempt to cover the most important branches of biomathematics that are treated by deterministic models, with the emphasis placed on supraindividual biology. Some branches were deliberately excluded, the most important of which, perhaps, deal with the functioning of the living body (models of the heart, neural networks, etc.). I believe that the modeling of these systems is very similar to the modeling of complex systems of technology and, as a consequence, the same ideas apply for the most part.

The first two Chapters deal with population dynamics. The first one shows models in which time is passing in discrete steps; these are applied, for example, to modeling the dynamics of insect populations. In this Chapter I also treat a model of a single population with age structure. In the Chapter 2 time is a continuous variable, and different relations of species interaction, predator-prey, competition, cooperation etc. are examined. In addition, models are shown that take into consideration the delay in biological systems, age structure, and spatial distribution. Chapter 3 explores the spread of epidemics using the simplest Susceptibles-Infectives-Removed model, a model for sexually transmitted diseases, and one for pair formation. The spread of epidemics in a spatial domain is also presented. Chapter 4 on evolution explores the fundamental principles of population genetics, and the problems of evolution leading to the appearance of living creatures and immunology. Chapter 5 gives an overview of René Thom's elementary catastrophe theory whose aim was to model the evolution of the embryo and treats models of pattern formation that explain animal coat markings in Nature. In Appendix 1 the tools from linear algebra, difference equations, and stochastic processes that are needed are presented and some treatment of chaotic dynamics is included. Appendix 2 deals with ordinary differential equations the emphasis is on stability and bifurcations. Appendix 3 treats partial differential equations with an emphasis on reaction-diffusion equations and Turing instability. Appendix 4 contains a light introduction into local Riemannian geometry, with the purpose of explaining the metric in the phase space of genotype frequencies where the equations of selection govern the dynamics.

Some of the figures were prepared with MAPLE-V and some with PHASER (Kocak [1989]); this is noted in the figure captions.

I hope that this book (having been kept so slender with considerable effort) may serve as a good introduction into this fascinating subject and will invoke interest which then may lead to more in-depth studies and research. I believe also that it may be used as a textbook for an introductory graduate course.

In the preparation of this volume Tamás Czárán's comments were extremely useful. I have learned much on immunology in Gábor Tusnády's seminar. Gábor Salfer helped to solve the problems that arose in my struggle with computers. Flora Géczy did an excellent job in preparing the final latex copy and producing some of the figures. I thank them all. I also thank my wife Kati for her patience and understanding because the larger part of the time spent writing was robbed from her.

My work on this book was partially supported by the Hungarian National Foundation for Scientific Research grant No. T029893 and T031716.

Medellín, February 2001 Miklós Farkas

Contents

Chapter 1

DISCRETE POPULATION MODELS

Population dynamics looks at the problem of how the number, the quantity of a well-defined group of living creatures, a species or a system of species, that is, those that share a common habitat, varies in time. Living creatures are born, reproduce, and die at a certain rate that depends on circumstances, including their specific genetically determined properties, the quantity of food available, their own density etc., and in case of a shared habitat, on the properties of those species with whom they live together. In this chapter we deal first with species of *nonoverlapping generations*. This means that the parent generation has disappeared by the time the next generation is born. One may imagine some insects that lay their eggs in the soil in the autumn and then die while the next generation is born the next spring. Next we consider a single species with discrete age groups.

1.1 Nonoverlapping Generations and Discrete Time Models

In this Section a single isolated population will be considered first. Its number or abundance at time t is denoted by N_t. Time is measured in discrete units (seconds, hours, years etc.) and it is assumed that the number of the generation of the moment (year etc.) t determines the number of the next generation, that is, the number N_{t+1}. In other words, this means that the previous generations influence the abundance of the generation at time $t+1$ only through the generation at time t. It is also assumed that the circumstances that may have an effect on reproduction, food, temperature etc. remain the same, for example, each year is like the previous one. Consider the difference between the numbers of the $(t+1)$st and the tth generation. If we divide this difference by the quantity of the tth generation we obtain the *per capita growth rate at time*

t. It is usually given in percentages. Population dynamics depends on how this per capita growth rate at time t depends on the actual size of the population. The simplest assumption is that this rate is constant. If this constant is negative then this means that there are fewer in each successive generation. If this negative rate is constant, the obvious consequence is a population that dies out rapidly. If this constant is positive then the equation that governs the dynamics is

$$(N_{t+1} - N_t) / N_t = r ,$$

where the constant $r > 0$ is now the per capita growth rate of the population. This equation can be written in the form

$$N_{t+1} = (1 + r) N_t . \tag{1.1.1}$$

If we express the number at time $t+2$ by the number at time $t+1$, and then the number at time $t + 3$ by the number at time $t + 2$ and so on, then the number of the generation at time $t + n$ will be

$$N_{t+n} = (1 + r)^n N_t .$$

As $r > 0$, this clearly means that the numbers go to infinity as time increases indefinitely. If the per capita growth rate is, for example, 2%, then the hundredth generation numbers $1.02^{100} = 7.24$ times as much as the original one. In Nature such *exponential growth* cannot go on indefinitely because some limiting factor of the environment, lack of food, oxygen, space etc. or simply the adverse effects of overcrowding, slows down growth sooner or later. We arrive at a more realistic model if we assume that the per capita growth rate is a decreasing function of the abundance of the population, which equals zero when the size of the population reaches the maximum that can be maintained by the environment. The simplest way to do this is to set the per capita growth rate as a linear function of the quantity with negative slope. In a graph of this function, the point where this line intersects the horizontal axis of the quantity is the maximum amount the environment can maintain. This value is called the *carrying capacity* and is denoted by K> 0. Accordingly, Eq. (1.1.1) is modified to

$$\begin{aligned} [N_{t+1} - N_t) / N_t &= r (1 - N_t / K) \quad \text{or} \\ N_{t+1} &= N_t (1 + r - rN_t / K] . \end{aligned} \tag{1.1.2}$$

Here $r > 0$ is called the *intrinsic growth* rate of the population. It prevails if N_t is small; then the per capita growth rate is approximately equal to r. If we look at Eq. (1.1.2) we see that in case N_t is less than the carrying capacity K then N_{t+1} will be larger than N_t, while if N_t is larger than K then N_{t+1} will be smaller than N_t. If N_t is equal to K then N_{t+1} will be the same. The variation of the size of the population according to Eq. (1.1.2) is called *logistic dynamics.* Besides $N = 0$ (when there is no population present), $N = K$ is its *equilibrium point.* For certain values of the intrinsic growth rate this point is stable in the

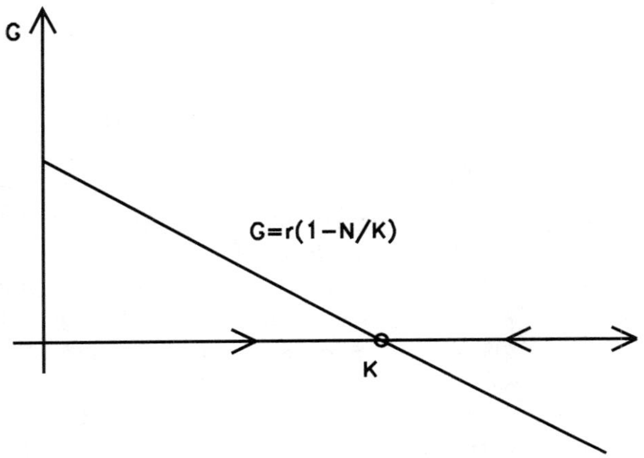

Figure 1.1.1: The growth rate of the logistic dynamics.

sense that if the population is higher or lower than this value its size goes to K (see Fig. 1.1.1).

Logistic dynamics has the great advantage that it does not let a population grow indefinitely, and if the population follows this rule then it settles down in the long run at a constant value, its carrying capacity. However, Eq. (1.1.2) has the disadvantage that if a very large value is substituted for N_t then N_{t+1} may be negative, which is meaningless. This difficulty can be overcome by the application of *exponential dynamics*:

$$(N_{t+1} - N_t)/N_t = e^{r(1-N_t/K)} - 1\,, \quad \text{or} \quad N_{t+1} = N_t e^{r(1-N_t/K)}\,, \quad r, K > 0\,. \tag{1.1.3}$$

Here again, if the size of the population is $< K$ then the next generation will be larger than the previous one and if the size is larger than the carrying capacity then the next generation will be smaller, and the population may finally settle down at K. One may substitute any positive number for N_t and the size N_{t+1} of the next generation will always be positive.

In the three cases discussed in the preceding, the set up is as follows. A function $F(N) = N \cdot f(N)$ is given such that if we divide it by N then we obtain the ratio of the size of the next generation to the actual one: $F(N)/N = f(N)$. In the first case the latter is constant, in the second it is a linearly decreasing function, and in the third it is exponentially decreasing. The dynamics starts at a certain time t, which will be taken as 0 in what follows and an initial size of the population N_0 is given. Then the size of the next generation is given by $N_1 = F(N_0) = N_0 f(N_0)$. The process continues like this. We have arrived at the concept of the *one-dimensional discrete forward dynamical system or semiflow* that is given by

$$N_{k+1} = F(N_k), \qquad (k = 0, 1, 2, 3, ...) . \tag{1.1.4}$$

If we substitute successively the obtained values of N into Eq. (1.1.4) we get a sequence $N_0, N_1, N_2, ..., N_k, \ldots$, which is called the path of N_0. It is also the path of any term N_k in it provided that we start the count of time at $t = k$. It may happen that a number E is such that the value of F at E is E, that is, $E = F(E)$. In this case E is an *equilibrium* (or *fixed*) *point* of the dynamical system, its path is $E, E, E, \ldots$. We say that the equilibrium point is *stable in the Lyapunov sense* if for any N_0 that is sufficiently close to E the path of N_0 stays near to E in the whole future. We say that E is *asymptotically stable* if it is stable in the Lyapunov sense and for any N_0 that is sufficiently close to E the path of N_0 tends to E as time tends to infinity. In Fig. 1.1.2 we show the graph of a function F, actually the one on the right-hand side of Eq. (1.1.3), the iteration process (1.1.4) and an asymptotically stable equilibrium.

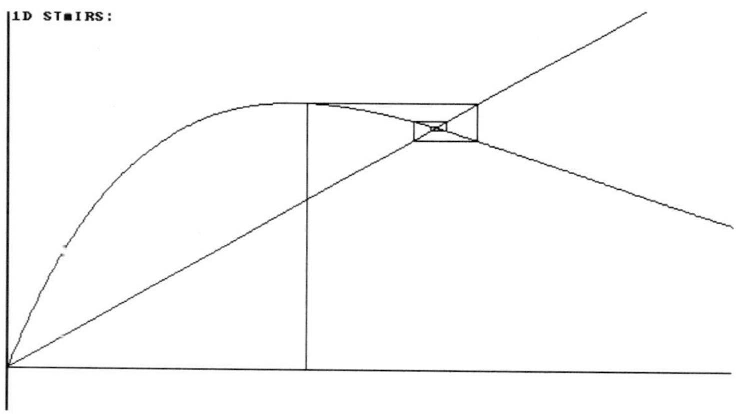

Figure 1.1.2: The dynamical system (1.1.3) with $r = 1.5$ and with $K = 1 = E$ an asymptotically stable equilibrium (PHASER).

It may happen that a point P is not an *equilibrium* but after n iterates of F one arrives back at P (and n is the least integer for which this happens); this means that

$$P = F(F(\ldots F(P))) = F \circ F \circ \ldots \circ F(P) = F^n(P) . \tag{1.1.5}$$

If this is the case we say that P is a *periodic point of period* n, and its path is then a periodic or closed path that closes in after n steps. Naturally, each point $P_k = F^k(P)$, $(k = 1, 2, 3, \ldots, n; P = P_0 = P_n)$ of the path is also a periodic point of period n. An equilibrium point is a periodic point of period 1. We say that the *periodic path is stable in the Lyapunov sense, respectively, asymptotically stable* if P is a Lyapunov stable, respectively, asymptotically

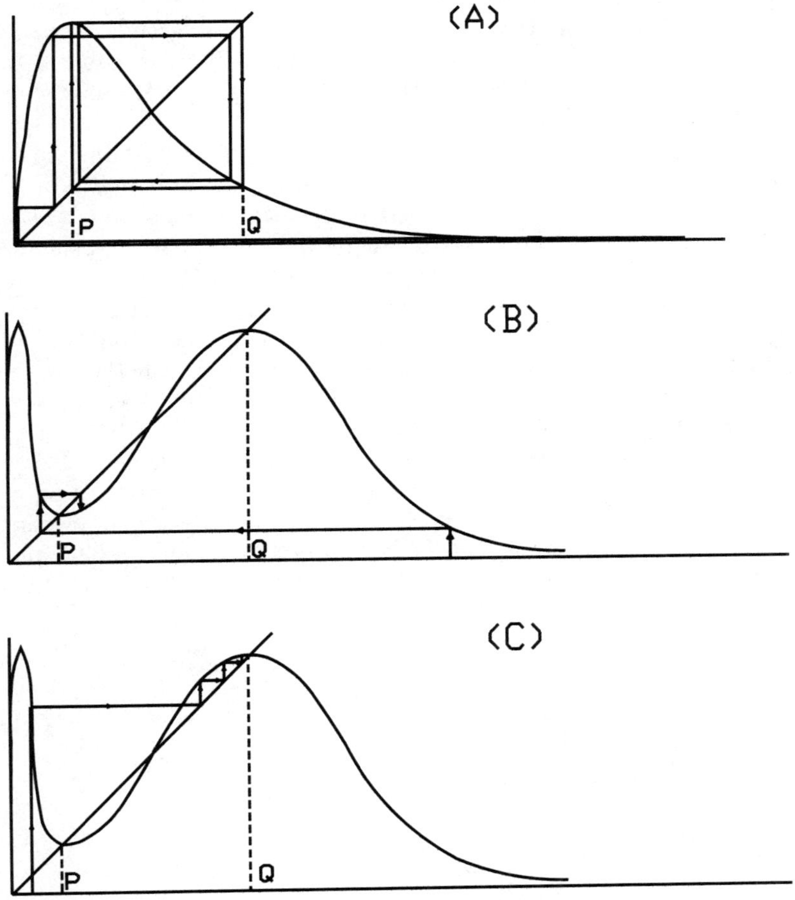

Figure 1.1.3: (A): The graph of $F(N) = Ne^{r(1-N/K)}$ with $r = 2.3$, $K = 1$, an asymptotically stable 2-periodic path P, Q. (B),(C): The graph of F^2 with asymptotically stable equilibria P and Q (PHASER).

stable equilibrium point of the dynamics defined by function F^n. That it is an equilibrium of F^n is clear from Eq. (1.1.5). If the period is 2, say, then this means that F maps P into a point Q and Q into P. Then $F(Q) = F(F(P)) = P$, and also $F(P) = F(F(Q)) = Q$. When the closed path P, Q is asymptotically stable then each path starting near to it tends to it, and each path of the dynamics determined by F^2 and starting near to P or Q, respectively, tends to P or Q, respectively. Figure 1.1.3 A shows the graph of the right-hand side of Eq. (1.1.3) with $K = 1$ and $r = 2.3$. In this case the dynamics has an asymptotically stable periodic path of period 2. Figure 1.1.3 B show the graph of the second iterate of this function with its two asymptotically stable equilibrium points, which form the graph of the 2-periodic path on (A).

When one observes that an insect population is larger every second year and smaller every first one then it may seem to be reasonable to model its dynamics by a one-dimensional (1D) discrete semiflow with an asymptotically stable 2-periodic path.

In Appendix 1.2 the 1D discrete forward dynamical systems are treated in some detail mainly because, besides their stable equilibria and periodic paths, they may present *chaotic dynamics* that, in such systems, can be studied relatively easily. Chaotic dynamics is briefly described in the appendix and how it may arise by an infinite sequence of *period doubling bifurcations* is discussed. Chaotic dynamics is now the focus in many branches of science. For example, if we have an insect population (whose subsequent generations appear yearly) that achieves a maximum population size every fourth year then we may try to describe its dynamics by a forward dynamical system that has a stable period-4 trajectory. If in case of an insect population one observes that the numbers (quantities) of subsequent populations vary chaotically, then one may draw the conclusion that the dynamics of the population may be modeled by a semiflow in the domain of the parameter where its dynamics is chaotic. Figure 1.1.4 shows the stable 4-periodic path of system (1.1.3) when $r = 2.65$ and $K = 1$; Fig. 1.1.5 shows the stable 8-periodic path when $r = 2.67$ and $K = 1$; and Fig. 1.1.6 shows the chaotic stage when $r = 3.3$ and $K = 1$.

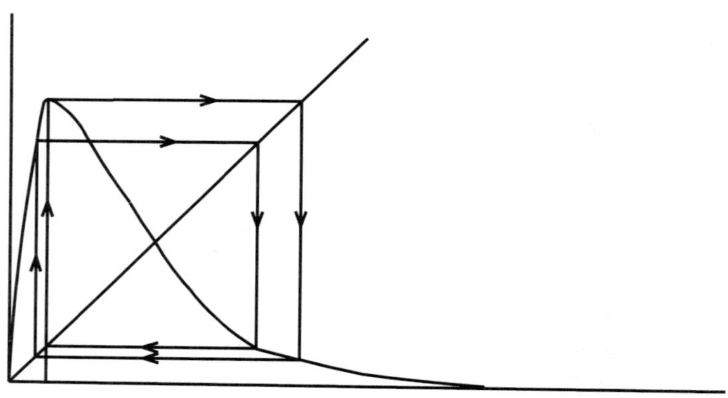

Figure 1.1.4: The stable 4-periodic path of system (1.1.3) at $r = 2.65$ and $K = 1$ (PHASER).

Discrete forward dynamical systems also can be applied in describing the dynamics of an ecological community consisting of interacting populations with nonoverlaping generations. In such a situation we speak of higher dimensional discrete semiflows. Here we are to treat a two-dimensional (2D) predator-prey system denoting the quantity of prey and predator at time $t = k$ ($k = 0, 1, 2, 3, \ldots$) by N_k and P_k, respectively. In what follows we shall speak about the *relative growth* N_{k+1}/N_k or P_{k+1}/P_k of the respective species instead

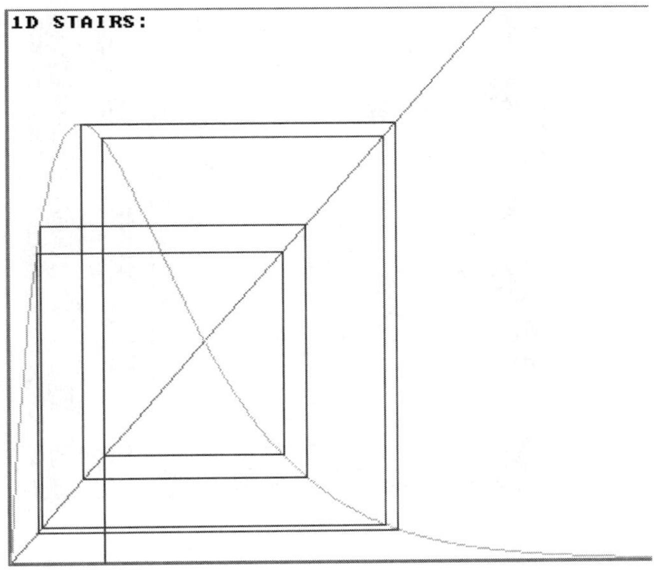

Figure 1.1.5: The stable 8-periodic path of system (1.1.3) at $r = 2.67$ and $K = 1$ (PHASER).

of the per capita growth rate. We want to have a model describing the interaction of the two species with the property that if there is no predation, $P = 0$, the system should reduce to the exponential dynamics of system (1.1.3) with respect to the prey and if there are predators present then they should reduce the relative growth of prey by a factor e^{-aP} where a is a positive number, so that e^{-aP} is < 1. Further, in case of the predator we suppose that its relative growth is proportional to the quantity of prey (food) available but the factor of proportionality is decreasing with the quantity of predator due to the, so called *intraspecific competition* in the predator population. An often studied model with these basic properties is

$$N_{k+1} = N_k e^{r(1-N_k/K-P_k a/r)}, \quad P_{k+1} = N_k \left(1 - e^{-aP_k}\right), \tag{1.1.6}$$

where $a > 0$ has already been described: e^{-a} is a kind of *predation rate*, where one unit of predator decreases the relative growth of prey by this factor; K>0 is the carrying capacity as in the preceding: if there is no predation, prey can grow in numbers up to this value in the long run (if there are predators present then, clearly, the exponential factor in the prey equation, that is, the relative growth of prey becomes < 1 before N can reach the value of the carrying capacity, in other words, prey begins to decrease earlier); $r > 0$ is again the intrinsic growth rate of prey: if the quantities of prey and predator are small (and also r is small), then the exponential factor is $\approx 1 + r$, which may be called the intrinsic relative growth because, in this case $N_{k+1}/N_k \approx e^r \approx 1 + r$. If we look now

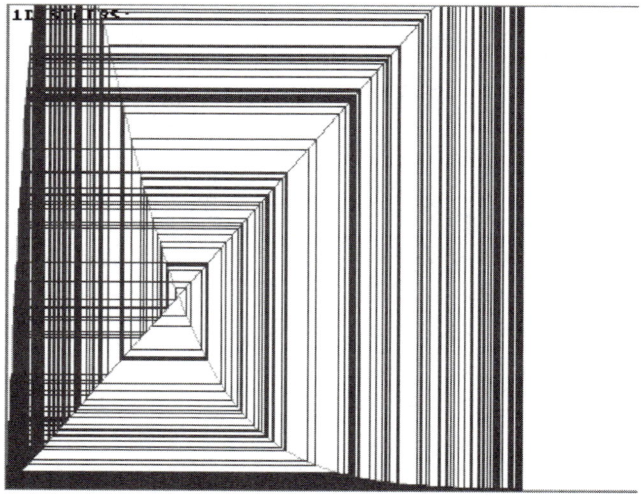

Figure 1.1.6: The chaos of system (1.1.3) at $r = 3.3$ and $K = 1$ (PHASER).

at the predator equation and divide both sides by P_k we see that the relative growth of the predator is given by the function

$$g(N, P) = N\left(1 - e^{-aP}\right)/P\,. \tag{1.1.7}$$

Thus, the relative growth is proportional to the quantity of prey but the factor of proportionality $\left(1 - e^{-aP}\right)/P$ is a decreasing positive function of the predator quantity, its limit at $P = 0$ is a, and its limit, as P tends to infinity, is zero. The dynamics of the predator-prey system (1.1.6) is generated by the pair of functions

$$(F(N, P), G(N, P)) = (Nf(N, P), Pg(N, P))\,,$$

where $f(N, P) = e^{r(1-N/K-Pa/r)}$ and the function g is given by Eq. (1.1.7). We discuss now the dynamics of this 2D semiflow studied by Beddington et al. (1975); background material is contained in Appendix 1.2. Here we mention only that *equilibria, periodic trajectories, and stability* can be defined in complete analogy to how this was done in the case of 1D semiflows. First it is to be noted, that system (1.1.6) inherited the property of the 1D exponential dynamics that if starting from a positive (N, P) the trajectories stay positive, therefore, in what follows the nonnegativity of N and P will always be tacitly assumed. If there are no predators, that is, we start from a point $(N, 0)$, then the trajectory stays on the N axis. If there is no food, that is, we start from a point $(0, P)$, then the next and all subsequent generations of the predator will be extinct and the predator dies out.

The equilibria of the system will be determined next. It is easy to see that $(0, 0)$ and $(K, 0)$ are fixed points. In order to find any equilibria in the interior of the positive quadrant of the plane N, P we divide the first equation of (1.1.6) by N_k, the second one by P_k and make the right-hand sides equal to one (the requirement is that $N_{k+1}/N_k = P_{k+1}/P_k = 1$). The system of equations that is to be satisfied by the coordinates is then

$$e^{r(1-N/K-Pa/r)} = 1\,, \quad N\left(1-e^{-aP}\right)/P = 1\,,$$

or

$$N/K + Pa/r = 1\,, \quad N = P/\left(1-e^{-aP}\right)\,. \tag{1.1.8}$$

The straight line determined by the first equation and the graph of the second equation can be easily drawn; see Fig. 1.1.7. The figure shows that there is an intersection in the positive quadrant only if $1/a < K$, that is,

$$aK > 1\,. \tag{1.1.9}$$

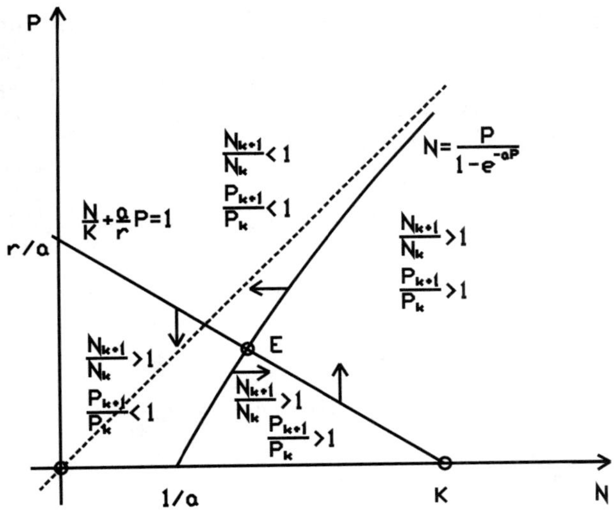

Figure 1.1.7: The phase portrait of system of equations (1.1.6) with the zero growth rate "isoclines" of prey and predator. The arrows show the direction in which the respective point moves.

This will be assumed in the sequel. It is also clear that if this condition holds then there is just one fixed point in the positive quadrant. The straight line and the other graph are important not only for finding the equilibrium point graphically, but they are the curves on which the relative growth of prey, respectively, predator is one. Above the straight line the relative growth of prey is < 1, that is, the prey is decreasing; below the line the relative growth is > 1, that is, the number of prey is increasing. This is intuitively reasonable; above the line there are too many predators or too many prey, resulting in a saturating effect in view of the intraspecific competition within the prey species. Above (and to the left of) the other graph the relative growth of predators is < 1, with predators decreasing because there are too many of them for the available food. To the right of (and below) this graph there is an abundance of food and relatively little competition within the predator species, resulting in a relative growth > 1. The stability analysis of the fixed points is carried out in Appendix 1.3. One has to linearize the system at the respective fixed point, that is, generate the Jacobi matrix of functions F and G and check the location of the eigenvalues. If the moduli of the eigenvalues are < 1, then the fixed point is asymptotically stable. It turns out that for any feasible choice of the parameters of the system the equilibria $(0,0)$ and $(K,0)$ are unstable. The equilibrium point inside the positive quadrant (denoted by E in Fig. 1.1.7) is asymptotically stable for relatively small values of the intrinsic growth rate r of the prey but as r is increased it loses its stability and at certain values of the parameters a and K the system may have periodic trajectories of high periods and behave chaotically in a bounded domain of the phase plane N, P. Figure 1.1.8 shows four different ways this system may behave at various choices of the parameters: (A) it has an asymptotically stable equilibrium point; (B) there is a closed invariant curve that attracts the trajectories but the motion on the curve itself seems to be irregular, that is, "chaotic"; (C) it has an asymptotically stable periodic trajectory of period 20; and (D) there is a bounded region of sharp contour in the phase plane inside which the dynamics seems to be chaotic.

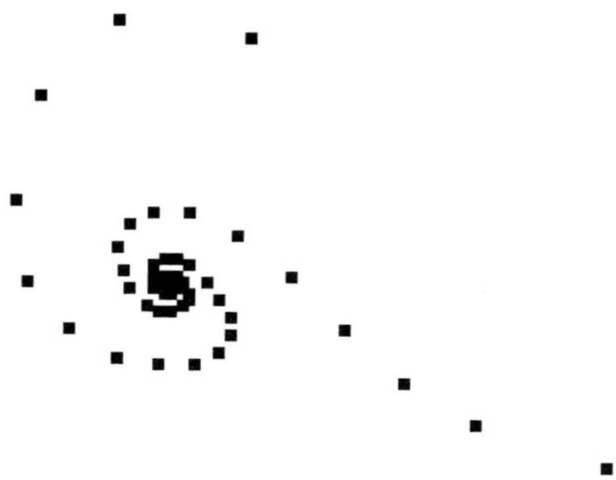

(A)

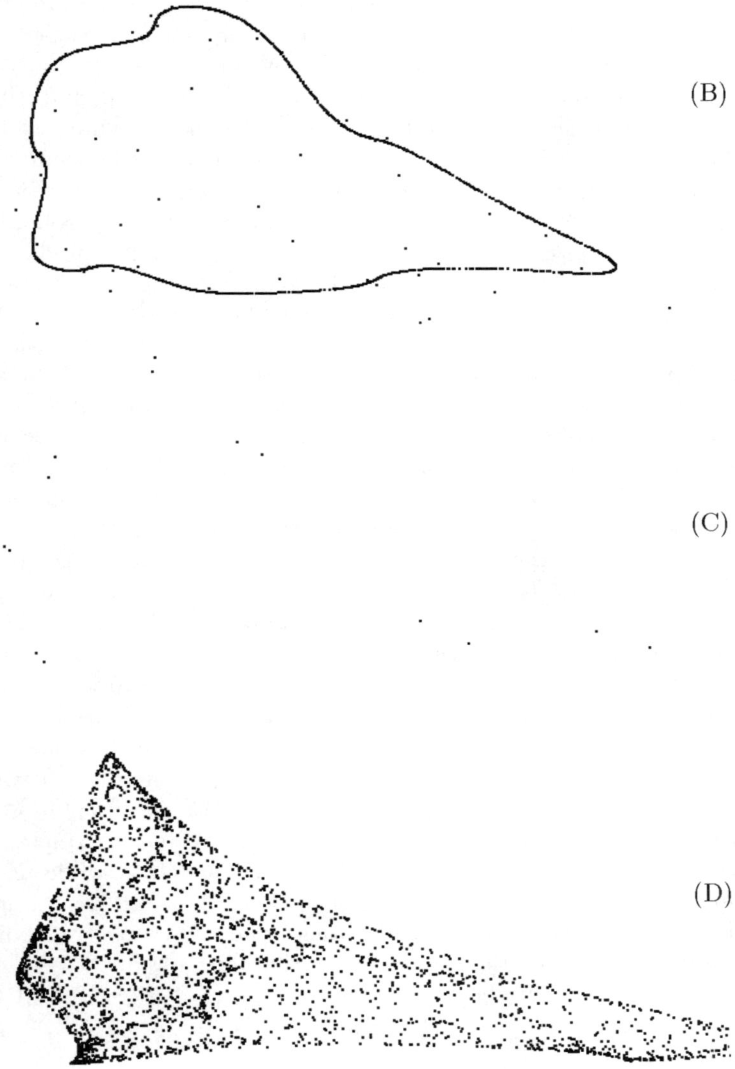

Figure 1.1.8: (A) Asymptotically stable fixed point of flow (1.1.6), $K = 1$, $a = 1.5$, $r = 0.5$; (B) attractive closed curve, $K = 10$, $a = 0.42998$, $r = 2$; (C) attractive periodic path of period 20 $K = 10$, $a = 0.48203$, $r = 2.4880$; and (D) chaotic attractor, $K = 10$, $a = 0.51178$, $r = 2.75$ (PHASER).

1.2 Population with Age Structure

Discrete flows or semiflows are also applied extensively in modeling other phenomena of the time evolution of populations. An important field where this happens is the dynamics of a single species in which age structure is to be taken into account. In the case of most long-lived organisms, the per capita birth rate is related to the total population but it is clear that the only age groups participating in reproduction are those that have reached maturity and thus if the number of newly born offspring is related to the number of those who produce them one will necessarily obtain different ratios depending on the age group considered. Similarly, death rate usually is also age dependent. There is another problem (which will not be considered here): the role played in the division of total population into two sexes for species that reproduce in a sexual way. We mention only that one generally gets more realistic results if the number of offspring is related to the number of females, the latter being the limiting factor in the reproduction (males are present mostly in superfluous numbers in nature). The dynamics of a population with age structure and the dynamics of the age distribution itself are also central problems in *human demography*. For this and other related problems see Keyfitz (1985). While there were several predecessors, an important method of handling this problem is attributed to Leslie (1945; 1948). We present here a concise treatment of the essence of the method; a detailed study can be found in Svirezhev-Logofet (1983).

The population (the female population) is divided into n age groups (for the human population a lifespan of 80 years is broken down into 5-yr groups, i.e., $n = 16$ age groups). The population is supposed to be closed, which means that newcomers come into the population only by birth and, naturally, into the first age group (for humans the group ranges from 0 to 4 years old) and members leave the population only by death. The contribution to the increase of age group 1 by the kth age group in unit time, which is now the span of the age groups (5-yr groups for humans) is $b_k x_k(j)$ where $x_k(j)$ is the number of individuals in the kth group at time j and b_k is their *fertility*, that is, the per capita offspring rate of those in this age group in unit time. (In order to illustrate the problems that arise it is to be noted that while in the case of humans $b_1 = b_2 = 0$, the fertility b_3 of those in the third group, ages from 10 to 14, is not necessarily zero because those who are 14 at the beginning of the time unit will be 19 by the end). Denote the *mortality* of those in age group k by d_k; this means that the percentage of those in age group k at time j who survive to form age group $k+1$ at time $j+1$ is $0 \le s_k = 1 - d_k \le 1$. Stated in formulaic terms,

$$x_1(j+1) = \sum_{k=1}^{n} b_k x_k(j) \quad \text{and} \quad x_{k+1}(j+1) = s_k x_k(j) \, ,$$

$$k = 1, 2, \ldots, n\, ; \quad j = 0, 1, 2, \ldots \, .$$

Denote the column vector of the numbers in the different age groups at time j

by $x(j) = \text{col}\,(x_1(j), x_2(j), \dots, x_n(j))$ and introduce the so-called *Leslie matrix*:

$$L = \begin{bmatrix} b_1 & b_2 & b_3 & \dots & b_n \\ s_1 & 0 & 0 & \dots & 0 \\ 0 & s_2 & 0 & \dots & 0 \\ . & . & . & \dots & 0 \\ 0 & 0 & \dots & s_{n-1} & 0 \end{bmatrix};$$

then, clearly,

$$x\,(j+1) = Lx\,(j)\,, \qquad j = 0, 1, 2, \dots \tag{1.2.1}$$

yields the dynamics of the age distribution of the population. In Eq. (1.2.1), the linear function Lx generates a flow if the Leslie matrix is regular, that is, $\det L \neq 0$. (Clearly, L is the Jacobi matrix of the flow in every point.) It is easily seen that $\det L = (-1)^{n+1}\, b_n s_1 s_2 \dots s_{n-1}$, so if the survivorships s and the fertility of the last age group are nonzero then Eq. (1.2.1) is a flow indeed. This is important because if L is singular its determinant is zero and then Eq. (1.2.1) is only a semiflow, which means that we are able to forecast how the age distribution of the population will look in the future but we cannot tell anything about the population's past. There is a tacit assumption made in this model, namely, that fertilities and survivorships are constant in time. This assumption has far-reaching consequences and is the starting point of the so-called *stable theory* of demography (however, we cannot go into it here). Eq. (1.2.1) implies that if the age distribution of the population is known at the initial time $j = 0$, then the age distribution at time j will be $x(j) = L^j x(0)$ where L^j is just the jth power of the matrix ($j = 0, 1, 2, \dots$). If Eq. (1.2.1) is a flow then if the distribution is known at time j we are able to determine the distribution at the initial moment $j = 0$, say, by $x(0) = L^{-j} x(j)$. It is natural to suppose that all survivorships s are positive, so that the regularity of the Leslie matrix depends on whether the fertility b_n of the last age group is positive. This is usually assumed in such a way that in the model those groups beyond reproductive age, the so-called *postfertility groups*, are not considered. These groups would affect the dynamics only if they competed within the population for some resource that is restricted. (In the human population this may mean that 50 years would be considered the life span, so that there would be only 10 age groups of 5-yr span each. To *disregard* older people in the *model* is more civilized than to *discard* them in real life as was the case, by necessity, in past ages when older members were led up a hill and abandoned or left with some food in the forest when the tribe moved on to its winter camp). Keyfitz (1985) shows, indeed, that while death rates due to cardiovascular and renal diseases were 15 times higher than the death rate due to certain diseases of infancy in the United States in 1964, eradication of heart disease and renal failures would have $\frac{1}{9}$ of the effect of eradication of diseases among infants because the first acts mainly upon groups near or in the postfertility age groups. He shows similarly that although death rates due to malaria and heart diseases is more or less the same in some tropical countries, eradication of the first would have a much greater effect on population growth than that of the second one. These observations show that variations of survivorships in postfertility age groups have virtually no or very little effect upon reproduction of the population.

All Leslie matrix entries are nonnegative. The most important simple facts about nonnegative matrices are summed up in Appendix 1.1. First of all it is obvious that starting from a positive initial vector, $x(0) > 0$ (meaning that every coordinate is positive), the corresponding trajectory stays in the positive orthant of the space: $Lx(0) > 0$. The Leslie matrix with the assumption that the fertility of the last age group is positive, $b_n > 0$, is indecomposable and, as a consequence, it has a positive eigenvalue $\lambda_1 > 0$ with an eigenvector $u = \text{col}\,(u_1, u_2, \ldots, u_n)$ with positive cordinates $u_k > 0$, $k = 1, 2, \ldots, n$ such that all other eigenvalues are in modulus less than or equal to λ_1. From now on we shall assume that L has n simple eigenvalues and there are no other eigenvalues with modulus equal to λ_1. The first assumption does not restrict generality considerably because by arbitrarily small changes of the entries of a given matrix we may obtain a matrix with this property–matrices without multiple eigenvalues form an "open and dense subset of the set of matrices". With respect to the second assumption, one can prove that it holds if either the fertility b_1 of the first age group is different from zero or there are two consecutive fertilities b_j and b_{j+1} that are nonzero, which is a fairly reasonable supposition. Summing up our assumptions the characteristic equation of the Leslie matrix is

$$\det(L - \lambda I) = (-1)^n \left(\lambda^n - b_1\lambda^{n-1} - b_2 s_1 \lambda^{n-2}\right.$$

$$\left. -b_3 s_1 s_2 \lambda^{n-3} - \ldots - b_n s_1 s_2 \ldots s_{n-1}\right) = 0\,, \qquad (1.2.2)$$

and the eigenvalues $\lambda_1, \lambda_2, \ldots, \lambda_n$ are such that $\lambda_i \neq \lambda_j$ if $i \neq j$, $\lambda_1 > 0$, and $|\lambda_j| < \lambda_1$ for $j = 2, 3, \ldots, n$. The eigenvectors corresponding to the eigenvalues $\lambda_1, \lambda_2, \ldots, \lambda_n$ are denoted by $u, u^2, u^3, \ldots, u^n$, respectively. Any initial vector $x(0)$ can be expressed uniquely as a linear combination of these eigenvectors,

$$x(0) = \tilde{x}_1 u + \tilde{x}_2 u^2 + \ldots + \tilde{x}_n u^n\,,$$

where coefficients have been denoted by using a tilde because they are not the original coordinates of the vector $x(0)$ in the original natural basis of the space but are linear combinations of them. It is to be noted that in this decomposition $\tilde{x}_1 > 0$ because both $x(0)$ and u are positive vectors, and no other eigenvector u^k can be positive. Now, the flow takes this initial age distribution in time j into

$$x(j) = L^j x\,(0) = \tilde{x}_1 L^j u + \sum\nolimits_{k=2}^{n} \tilde{x}_k L^j u^k = \tilde{x}_1 \lambda_1^j u + \sum\nolimits_{k=2}^{n} \tilde{x}_k \lambda_k^j u^k\,. \qquad (1.2.3)$$

There are three possibilities for the long time behavior of the flow.

Case A. $\lambda_1 > 1$. In this case all of the trajectories (except $x = 0$) tend to infinity as time j tends to infinity because $\lambda_1^j \to \infty$ and this may be the case with some further members in the preceding sum:

$$\lim_{j \to \infty} L^j x\,(0) = \infty\,.$$

This means that the population is blowing up but after a long time (for large j) the sum on the right-hand side of Eq. (1.2.3) can be neglected compared to the first term, because λ_k^j, $(k = 2, 3, \ldots, n)$ are small with respect to λ_1^j, so that the long-term age structure of the population will be approximately similar to u but on a much larger scale. In reality we expect that after some time fertilities and survivorships will change.

Case B. $0 < \lambda_1 < 1$. In this case the modulus of the rest of the eigenvalues is also < 1, so that every member in the forementioned sum and, as a consequence, every trajectory tends to zero:

$$\lim_{j \to \infty} L^j x(0) = 0$$

for arbitrary initial $x(0)$. The equilibrium point $x = 0$ is *globally* asymptotically stable (see Theorem A1.3.1)–it attracts *all* of the trajectories. The population dies out.

Case C. $\lambda_1 = 1$. First, the flow has equilibria besides zero iff 1 is an eigenvalue because then $Lu = u$. In this case every scalar multiple of the vector u is also an equilibrium because of linearity. If we look at the characteristic equation and substitute 1 for λ we see that 1 is a root iff

$$b_1 + b_2 s_1 + b_3 s_1 s_2 + \ldots + b_n s_1 s_2 \ldots s_{n-1} = 1 .$$

Now the rest of the eigenvalues are in modulus < 1, the largest eigenvalue. This implies that as time goes to infinity every term in the sum on the right-hand side of Eq. (1.2.3) goes to zero. This means that

$$\lim_{j \to \infty} L^j x(0) = \tilde{x}_1 u , \tag{1.2.4}$$

which has important implications. In this case the parameters of the system, the fertilities, and the survivorships determine the long-term age distribution of the population given by the positive vector u. Regardless of which positive age distribution the population starts from, the limiting distribution will be a positive scalar multiple of u. The initial age distribution $x(0)$ determines the factor $\tilde{x}_1$ linearly; the population will be larger or smaller depending on how large $\tilde{x}_1$ is, but the ratios of the different age groups to each other will always be the same in the long run. It is also clear that if the initial distribution $x(0)$ is disturbed slightly, then $\tilde{x}_1$ will also vary only slightly, so that the final age distribution of the population will remain close to the previous one–all equilibria (positive constant times u) are stable in the Lyapunov sense (see Definition A1.2.2).

We note that Keyfitz (1985), working with Mexican data from 1966, found that $\lambda_1 = 1.1899$.

Chapter 2

POPULATION DYNAMICS IN CONTINUOUS TIME

In this chapter we treat species with *overlapping generations*, that is, species whose parents are normally present when the new generation is born (for example, mammals). In this case the dynamics is to be considered in continuous time. For the topics studied the main tools used are differential equations. For background material see Appendix 2.

2.1 The Logistic Equation and Human Demography

At the end of the eighteenth century Malthus (1798) published his *Essay on the Principle of Populations*, according to which the velocity of the growth of a population is proportional to the quantity of the population: twice as many members produce twice as many offspring in unit time. This is a very simple but quite reasonable assumption and can be expressed in the language of differential equations the following way. Denote the size (the quantity, the numbers) of the population at time t by $N(t)$. This function is considered to be a nonnegative function that varies continuously in time and may change its value by an arbitrary small quantity. Now, if we have in mind bacteria, grass, or sardines, then this is accebtable because we may measure them in milligrams or tons but if we talk about lions or human beings then this is just an approximation, and it may considered to be reasonable only if the numbers are large and we do not have to meditate over the result when the equations tell us that the population counts at a certain moment 1,932,194,502.13 members. Derivatives with respect to time will be denoted by dot, and the Malthusian population can be modeled then by the equation

$$\dot{N}(t) = rN(t) , \tag{2.1.1}$$

where $r > 0$ is a constant called the *intrinsic growth rate* of the population. The equation also can be put in the form

$$\dot{N}(t) / N(t) = r ,$$

which tells us that the *per capita growth rate* (the number of offspring produced by one member of the population in unit time the total number of offspring in unit time divided by the quantity of the population) is the constant r. In a natural population there is both an ongoing birth and death process. If migration is disregarded and the per capita birth and death rates are denoted by b and d, respectively, then the growth rate is $r = b - d$ and it is positive only if birth rate is higher than death rate.

The solutions of the differential equation (2.1.1) are $N(t) = N_0 e^{rt}$, where $N_0 = N(0) > 0$ is the size of the population at the initial time. Thus, a Malthusian population is growing exponentially. By observing the data and measuring how much time is needed for the doubling of a population, one may determine the intrinsic growth rate. Denote the doubling time by T; then $2N_0 = N_0 e^{rT}$, or dividing by N_0 and taking logarithms

$$r = (\ln 2) / T . \tag{2.1.2}$$

Let us try to apply this formula to the demographic history of the human race as shown by Keyfitz (1985). Taking into account the inexact knowledge we have now about *Australopithecus africanus*, *Homo sapiens neanderthalensis* and others, we may assume that just one couple of the *Homo sapiens sapiens* lived 1 million years ago (we may call them Adam and Eve). Now the descendants number approximately 6 billion. How many doublings were needed for that? The answer is we are now between the 32nd and 33rd doublings. Thirty two and a half doublings in 1 million years means that 31,000 yrs were needed on average for one doubling. By applying Eq. (2.1.2) we obtain for the intrinsic growth rate of the human race $r = (\ln 2)/31000 = 0.000022$. If we suppose that death rate d was on the average 4%, that is, $d = 0.040000$, then the birth rate must have been $b = 0.040022$. This means that during the past 1 million years the birth and death rates were extremely close to each other–while 40,000 out of 1 million people died yearly only 40,022 were born. However, if one looks back over the last 150 yrs. one finds that there were roughly two and a half doublings, with the last one taking ≈ 32 yrs. This extraordinary situation occurred because the almost complete eradication of several lethal diseases and the relative abundance of food in most parts of the world caused the death rate to fall to 2% and large parts of the world still maintain a birth rate of $\approx 4\%$. There are now some projections by certain teams of demographers (e.g., those at IIASA in Laxenburg, Austria) according to which there will be no more doublings of the human race and its size will settle at ≤ 12 billion by the end of the next century. We shall return to this point at the end of this Section.

Objectors to the Malthusian model said that nowhere in Nature is it possible to have unlimited exponential growth of any population over the long run. In the first half of the 19th century Verhulst modified the Malthusian model, contending that because some resource is available only in limited quantities, every population grows into a saturated phase from which it can not grow further; the ecological habitat of the population can *carry* just so much of it and not more. When this principle is put into the form of an equation it is assumed that the per capita growth rate is not a constant but a decreasing function of the size of the population, and reactes zero as the population achieves a size that can be maintained; further, any population reaching a size that is above this value will experience a negative growth rate. The simplest way to realize this is by assuming that the per capita growth rate is a decreasing linear function of the size:

$$\dot{N}(t)/N(t) = r(1 - N(t)/K) ,$$

where $K > 0$ is the *carrying capacity* of the habitat. Multiplying by $N(t)$ we obtain the *Verhulst* or *logistic differential equation*:

$$\dot{N} = rN\,(1 - N/K)\ . \tag{2.1.3}$$

Indeed, if $N > K$ then N is decreasing because its time derivative is negative and if $0 < N < K$ then N is increasing. The equation has two equilibria, 0 and K. The derivative of the right-hand side $F(N) = rN(1 - N/K)$ is $F'(N) = r(1 - 2N/K)$. At equilibrium point 0 we have $F'(0) = r > 0$, so that this equilibrium is unstable. At $N = K$ we have $F'(K) = -r < 0$, so that this equilibrium is asymptotically stable. As the differential equation is a separable one, it can be solved explicitly. The solution with initial value $N_0 = N\,(0) > 0$ is

$$N(t) = N_0 K e^{rt}/(N_0(e^{rt} - 1) + K)\ , \qquad t \geq 0\ .$$

From here it can be seen that every solution with positive initial value tends to K indeed as time tends to infinity; see Fig. 2.1.1. It is worthwhile to determine the second time derivative of N. This can be done easily using the differential equation satisfied by the function N:

$$\ddot{N} = F'\,(N)\,\dot{N} = r\,(1 - 2N/K)\,rN\,(1 - N/K) = r^2 N\,(1 - N/K)\,(1 - 2N/K)\ .$$

No solution except the constant solutions $N \equiv 0$ and $N \equiv K$ assumes the value 0, respectively K. Apart from these, the second time derivative of each solution with initial value $< K/2$ becomes zero only when the solution assumes the value $N = K/2$. This means that these solutions are convex down before they have reached $K/2$ and they are concave down above this or, in other words, the rate of increase $\dot{N}$ is increasing while the population size is below half the carrying capacity and it is decreasing above this value (those solutions that have started from initial values already $> K/2$ and $< K$ are concave down from start).

If we accepted the Verhulst model for the human race and observed that while the size is still increasing the rate of increase has started to decrease, then we could draw the consclusion that the population size has reached half its carrying capacity, from where the carrying capacity itself could be determined by multiplying the observed size by two. If this is now the case then the carrying capacity of the Earth is 12 billion human being. Naturally, I do not want to say that serious demographers use the Verhulst model to draw consequences. It is considered merely a "toy model" now but it still shows certain properties that are also present in much more sophisticated models.

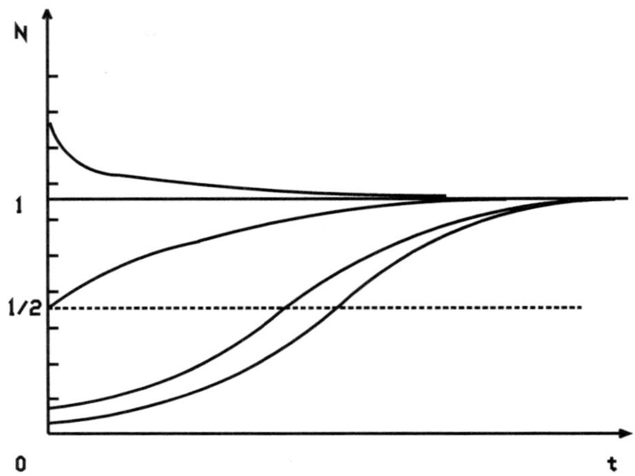

Figure 2.1.1: Solutions of Eq. (2.1.3), $r = 0.5$, $K = 1$; the solutions tend away from 0 and tend towards 1, points of inflexion are on the line $N = 1/2$ (PHASER).

2.2 Two Interacting Species–Predator and Prey

In a *habitat* (an ecologically and perhaps geographically distinguishable and well-defined place, a lake, a forest, an island, a tree, a test tube) where several species live together one may select two that seem to be the most important to each other and neglect the effect of the rest of the species in order to obtain a deeper insight into their interaction. Although the results obtained this way are abstract and thus may be erroneous, they may help in our understanding of the dynamics. If the two species chosen are not neutral to each other then there are, in principle, three ways for them to interact: (i) one of them may serve as food for the other; in which case we speak about a *predator-prey* interaction; (ii) the two species may compete for the same resources; in this case we speak about a *competitive system*; and (iii) they may help each other exist; then we say that they live in *symbiosis* or that they are *cooperative* or *mutualist*. Competitive and cooperative systems will be treated in the next Section. We turn now to predator-prey systems.

Modern mathematical population dynamics started with the works of Volterra (1931) and Lotka (1924). A system where the per capita growth rates of the participating species are linear functions of the quantities (densities) of the species is called a *Lotka-Volterra system of differential equations.* In our case here, we have a prey species whose quantity at time t will be denoted by $N(t)$ and a predator species whose quantity at time t will be denoted by $P(t)$. The per capita growth rate of the prey is

$$\dot{N}/N = \varepsilon\,(1 - N/K) - \alpha P\,,$$

where K, ε, α are positive constants, with K the carrying capacity of the environment with respect to the prey (if there is no predation, P is zero, then prey is following the logistic dynamics), ε the intrinsic growth rate of prey, and α the *predation rate* (one unit of predator decreases the growth rate of prey by this quantity). The per capita growth rate of the predator is

$$\dot{P}/P = -\gamma + \beta N - \delta P\,,$$

where γ and β are positive constants, and δ is a nonnegative constant, with γ the *intrinsic mortality* of the predator (if there is no food present, N is zero, and the predator quantity P is small then predator is dying out by this rate); however, if δ is nonzero then the predator's mortality is an increasing linear function of the quantity P due, perhaps, to other restrictive resources for which there is an *intraspecific competition* within the predator species; finally, β is the *conversion rate*, that is, the predator is decreasing the prey quantity in unit time by αNP and this increases the quantity of predators in unit time by $(\beta/\alpha)\alpha NP$ (normally $\beta/\alpha < 1$ because of some loss in the process of converting prey biomass into predator biomass). Multiplying by N and P, respectively, we obtain the general *2D Lotka-Volterra predator-prey system of differential equations*:

$$\dot{N} = N\varepsilon\,(1 - N/K) - \alpha NP\,, \qquad \dot{P} = P(-\gamma + \beta N - \delta P)$$

or

$$\dot{N} = N\,(\varepsilon - \varepsilon N/K - \alpha P)\,, \qquad \dot{P} = P\,(-\gamma - \delta P + \beta N)\,. \tag{2.2.1}$$

For this system the coordinate axes are *invariant.* If N was zero then it remains zero and so does P. Because trajectories of such autonomous systems can not cross each other (see Appendix 2.1), the positive quadrant of the N, P plane is also invariant–no trajectory can enter or leave this quadrant. We are interested only in the quadrant where the nonnegative values of N and P can be interpreted as population densities or quantities. The first question to be answered is, what are the equilibria in the positive quadrant? By making the right-hand sides equal to zero and solving the equations, we see that the only equilibria are $(0,0)$, $(K,0)$, and $E = (\overline{N}, \overline{P})$ where

$$\overline{N} = K\,(\varepsilon\delta + \alpha\gamma)\,/\,(\alpha\beta K + \varepsilon\delta)\,, \qquad \overline{P} = \varepsilon\,(\beta K - \gamma)\,/\,(\alpha\beta K + \varepsilon\delta)\,,$$

with the last one being in the interior of the positive quadrant only if

$$\gamma < \beta K. \tag{2.2.2}$$

We note that if $\gamma = \beta K$ then E coincides with $(K, 0)$. The next question to ask is, are these equilibria stable or not? To provide an answer to this question for the first two equilibria, we write out the Jacobi matrix of the right-hand sides and determine the eigenvalues. The Jacobi matrix is

$$J = \begin{bmatrix} \varepsilon(1 - 2N/K) - \alpha P & -\alpha N \\ \beta P & -\gamma + \beta N - 2\delta P \end{bmatrix}.$$

At $(N, P) = (0, 0)$ an eigenvalue is positive and the other one is negative, so that the origin is an unstable equilibrium point, more precisely a saddle. At $(N, P) = (K, 0)$ the characteristic polynomial is $\lambda^2 + \lambda(\varepsilon + \gamma - \beta K) + \varepsilon(\gamma - \beta K)$, so that by Theorem A1.1.2 if $\gamma > \beta K$ then this equilibrium point is asymptotically stable and if Eq. (2.2.2) holds then it is unstable. This condition of stability is intuitively clear. If predator mortality is high, the conversion rate is small (i.e., the predator is processing food inefficiently) and food is scarce (K is small), then the predator dies out; conversely, if predator mortality is low and both the conversion rate and carrying capacity are high, then the equilibrium representing the absence of the predators is unstable. Exactly in this last case do we have an equilibrium in the interior of the positive quadrant. Now we assume that Eq. (2.2.2) holds, so that the only possibly stable equilibrium is E. One could also attempt to determine its stability by finding the position of the eigenvalues but the calculations are too cumbersome and they would establish only local stability. We apply instead Lyapunov's direct method (see Appendix 2.2).

It is easy to see that $x - 1 > \ln x$ for $x > 0$, $x \neq 1$. Therefore we introduce the *Lyapunov function*

$$\begin{aligned} L(N, P) &= \overline{N}\left(N/\overline{N} - 1 - \ln\left(N/\overline{N}\right)\right) + (\alpha/\beta)\,\overline{P}\left(P/\overline{P} - 1 - \ln\left(P/\overline{P}\right)\right) \\ &= N - \overline{N} - \overline{N}\ln(N/\overline{N}) + (\alpha/\beta)(P - \overline{P} - \overline{P}\ln(P/\overline{P}))\,, \end{aligned}$$

which is positive for all (N, P) in the open positive quadrant except at point $E = (\overline{N}, \overline{P})$, where it is zero. The derivative of L with respect to the system (2.2.1) is obtained by a lengthy but straightforward calculation:

$$\dot{L}_{(2.2.1)}(N, P) = -\left(\left(N - \overline{N}\right)^2 \varepsilon/K + \left(P - \overline{P}\right)^2 \alpha\delta/\beta\right),$$

and this is negative definite with respect to the point E provided that $\delta > 0$. As a consequence, equilibrium point E is globally asymptotically stable with respect to the interior of the positive quadrant; all solutions with positive initial values tend towards this point as time tends to infinity. We note that the same Lyapunov function works for the equilibrium $(K, 0)$ in case $\gamma \geq \beta K$, that is, under this condition this equilibrium is not only locally but globally asymptotically stable with respect to the positive quadrant.

If $\delta = 0$, that is, the predator mortality is the constant γ, it does not depend on the quantity then the derivative of our Lyapunov function with respect to the system is $\dot{L}_{(2.2.1)}(N, P) = -\left(N - \overline{N}\right)^2 \varepsilon/K$, which is only positive semidefinite; it is zero at $\left(\overline{N}, P\right)$ for all $P > 0$. In this case $\overline{N} = \gamma/\beta$ and the vertical straight line $N = \gamma/\beta$ does not contain trajectories apart from the equilibrium $(\gamma/\beta,\ \overline{P})$. This implies, using by the Barbashin-Krasovskiy Theorem (see Appendix 2.2), that the equilibrium point is also globally asymptotically stable in this case. The phase portrait is shown for this case on Fig. 2.2.1.

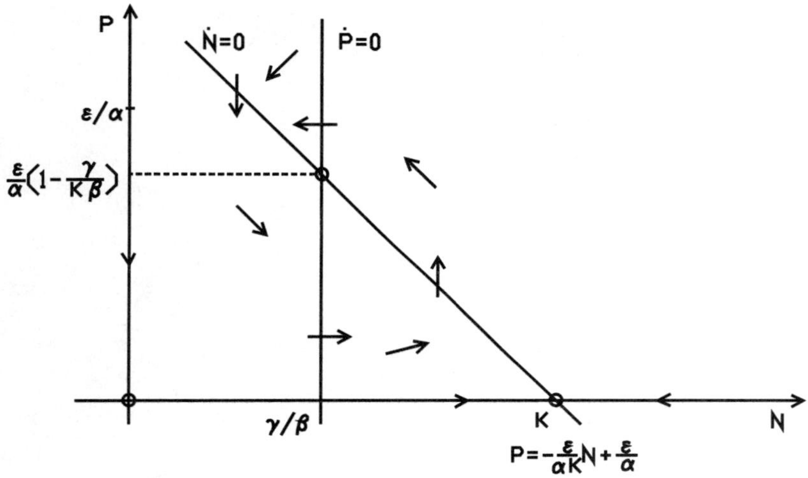

Figure 2.2.1: The phase portrait of system (2.2.1) with $\delta = 0$ under condition (2.2.2).

If addition to $\delta = 0$ the carrying capacity K can be considered infinite, or in other words, there is no intraspecific competition within the prey species, system (2.2.1) reduces to

$$\dot{N} = N(\varepsilon - \alpha P)\ , \qquad \dot{P} = P(-\gamma + \beta N)\,, \tag{2.2.3}$$

where $\varepsilon, \alpha, \gamma$, and β are positive constants (with the same meanings as before). This was the first model constructed by Volterra after World War I in order to explain why the quantity of predator fish increased relative to prey fish in the Adriatic Sea during the war when fishing had stopped because of naval battles between the Italian and Austro-Hungarian fleet. In this case the interior equilibrium is $(\overline{N}, \overline{P}) = (\gamma/\beta,\ \varepsilon/\alpha)$ and we may apply the same Lyapunov function as was previously used and which is now

$$\begin{aligned} L(N,P) &= N - \gamma/\beta - (\gamma/\beta)\ln(N\beta/\gamma) + (\alpha P/\beta - \varepsilon/\beta - (\varepsilon/\beta)\ln(\alpha P/\varepsilon)) \\ &= N - (\gamma/\beta)\ln N + (\alpha P/\beta) - (\varepsilon/\beta)\ln P + \text{const.} \\ &= (1/\beta)(\beta N - \gamma \ln N + \alpha P - \varepsilon \ln P) + \text{const.} \end{aligned}$$

Again a positive definite function (with the proper constant), it assumes its minimum at $(\overline{N}, \overline{P}) = (\gamma/\beta,\ \varepsilon/\alpha)$. Taking the derivative of L with respect to system, (2.2.3), we obtain $\dot{L}_{(1.4.3)}(N,P) \equiv 0$. This means that with a "negative semidefinite" derivative, the equilibrium is stable in the Lyapunov sense and further L is constant along the trajectories of the system, that is, it is a first integral. It can be shown that the level curves of this function, that is, the trajectories of the system, are closed curves around the interior equilibrium point and this means that this equilibrium is a center and every solution is periodic (see Fig. 2.2.2). Let us calculate the average values of N and P over a period along an arbitrary solution $(N(t), P(t))$. For this we divide the first equation by $N(t)$ and integrate over the interval $(0, T)$ where T is the period of this solution,

$$\int_0^T (\dot{N}(t)/N(t))\,\mathrm{d}t = \varepsilon T - \alpha \int_0^T P(t)\,\mathrm{d}t\,.$$

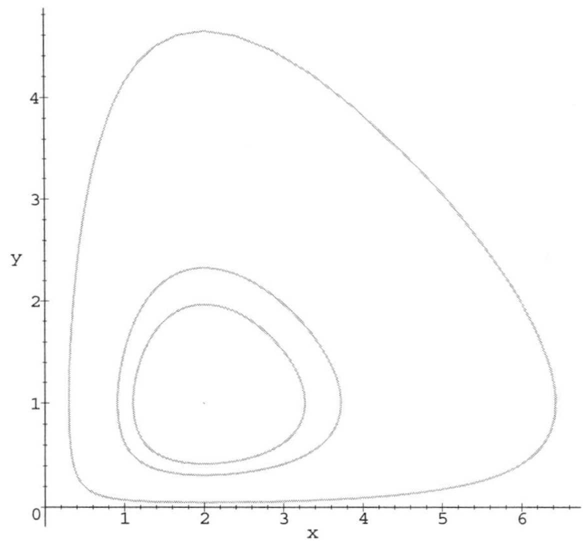

Figure 2.2.2: Phase portrait of system (2.2.3) (MAPLE).

The left-hand side is $\ln(N(T)) - \ln(N(0)) = 0$ because of the periodicity. Hence the average of P over a period is

$$(1/T)\int_0^T P(t)\,\mathrm{d}t = \varepsilon/\alpha\,.$$

Similarly, way we obtain

$$(1/T)\int_0^T N(t)\,\mathrm{d}t = \gamma/\beta\,.$$

Thus, no matter which solution we choose the averages are always the same and are equal to the respective values at equilibrium. If fishing is proportional to the quantities of the respective species of fish, then system (2.2.3) is replaced by

$$\dot{N} = N(\varepsilon - \alpha P) - \mu N, \qquad \dot{P} = P(-\gamma + \beta N) - \nu P,$$

where μ and ν are positive constants representing how catch decreases the per capita growth rate of the respective species. The last system is of the same structure as (2.2.3) where the only difference is that ε is replaced by $\varepsilon - \mu$ and γ is replaced by $\gamma + \nu$. As a consequence the averages are now $(\gamma + \nu)/\beta > \gamma/\beta$ for prey and $(\varepsilon - \mu)/\alpha < \varepsilon/\alpha$ for predator. From this it is clear that fishing at levels proportionate to quantities is advantageous for prey and disadvantageous for the predator and, naturally, any cessation in fishing has an inverse effect. The result is intuitively reasonable–although fishing decreases the growth rate of prey, at the same time it also reduces its enemy (it is to be assumed that there is no overfishing, i.e. the fishing rate is not larger than the intrinsic growth rate); from the point of view of the predator, fishing is doubly disadvantageous because it reduces its growth rate and the quantity of its food.

Often one may find it useful to generalize the Lotka-Volterra system in order to approximate the real situation better. The first step in this direction is to get rid of the assumption that per capita growth rates are linear functions of the quantities. Denoting again the quantity of prey and predator at time t by $N(t)$ and $P(t)$, respectively, we suppose that these functions satisfy the system of differential equations

$$\dot{N} = NF(N, P), \qquad \dot{P} = PG(N, P) \tag{2.2.4}$$

where F and G are continuously differentiable functions; prey can survive on its own, $F(0,0) > 0$, but there is a saturation effect acting when its quantity is high, so that there exists a $K > 0$ such that $F(K,0) = 0$ and in absence of predation the per capita growth rate $F(N,0)$ is positive or negative, respectively, for $N < K$ or $N > K$. Further, any predator quantity growth will decrease the growth rate of prey $F'_P(N,P) < 0$; the predator cannot survive without the prey $G(0,P) < 0$; the increase of prey quantity is beneficial to its growth rate $G'_N(N,P) > 0$; and there may be intraspecific predator competition $G'_P(N,P) \leq 0$. We also suppose that there is an equilibrium point for the system in the interior of the positive quadrant $E = (\overline{N}, \overline{P})$, $\overline{N}, \overline{P} > 0$,

$$F(\overline{N}, \overline{P}) = G(\overline{N}, \overline{P}) = 0.$$

All conditions imposed upon growth rates F and G are natural ones that acknowledge "N is the prey and P the predator." These conditions allow several kinds of prey and predator *zero isoclines*, that is, curves in the positive quadrant of the N, P plane where the respective growth rates are zero. The equations for the "prey curve" and "predator curve" are $F(N,P) = 0$ and $G(N,P) = 0$, respectively. A typical reasonable form of these curves applicable to most species is shown in Fig. 2.2.3a. Prey curve $F = 0$ has a maximum point somewhere between 0 and K. Below this curve the growth rate F is positive because prey is not yet overcrowded and there are only few predators are present, above this curve it is negative because the saturation effect acts and/or predation is strong. The interval from zero to the maximum point below the ascending branch of the prey curve is called the *Allée effect zone*–here prey is scarce and an increase in prey quantity is beneficial for the growth rate of prey. To keep the prey growth rate zero the increase of prey must be counterbalanced by an *increase* of predators. This is the case with most sexually reproducing species because if for example their density is too low, then males and females do not meet. At the descending branch of the prey curve where the effect of overcrowding is already felt, any further increase in prey must be counterbalanced by a *decrease* in predator quantity. The predator curve $G = 0$ is increasing and concave down, above and left to this curve the growth rate G of the predator is negative because there is not enough food and there are too many predators competing for food, below and right to this curve the growth rate is positive. The curve is ascending because growth in prey quantity makes growth of predator quantity possible, thus keeping the growth rate zero; and it is concave down because if prey is abundant then a further growth of prey has less effect on the growth of predators than is the case when food is scarce. Sometimes one may suppose that there is no intraspecific competition within the predator species, so that its growth rate depends only on the quantity of prey and not on its own quantity. In this case the predator curve is a vertical straight line like that seen in Fig. 2.2.3c. If the curves are such, then we have a positive equilibrium provided that the growth rate of the predator becomes positive at prey quantities that are less than the carrying capacity K. This last condition, namely, that the predator curve shall meet the axis N at a point left of K in order to have a positive equilibrium, corresponds to condition (2.2.2) for system (2.2.1).

Naturally, it may happen that the prey curve does not have an Allée effect zone but is descending right from the beginning. Such is the case with the Lotka-Volterra system (2.2.1). In this case the prey curve is a straight line with negative slope intersecting the axis P at ε/α and the axis N at K and the predator curve (if δ is nonzero) is another straight line with positive slope intersecting the axis P at $-\gamma/\delta$ and the axis N at γ/β (Fig. 2.2.4). As we have seen, if the two straight lines intersect each other in the interior of the positive quadrant (i.e., if condition (2.2.2) holds) then the point of intersection E is globally asymptotically stable with respect to the interior of the positive quadrant. What is the stability situation for the general predator-prey system (2.2.4)?

There is an intuitive way of seeing what we may expect. We call this the *method of the little arrows* (see Farkas and Freedman 1989 a and b). We draw pictures of prey and predator curves and look at the point of intersection, which is our equilibrium. The system moves from equilibrium in four special ways–by increasing or decreasing either the quantity of prey or of predator. We then check what the dynamics of the system does to this perturbed position. The signs of the coordinates of the vector field of the system, that is, the signs of the functions $F(N, P)$ and $G(N, P)$, show qualitatively the direction of the field and by that they also show whether the dynamics tries to decrease or increase the distance from equilibrium E. If in the majority of cases the dynamics tries

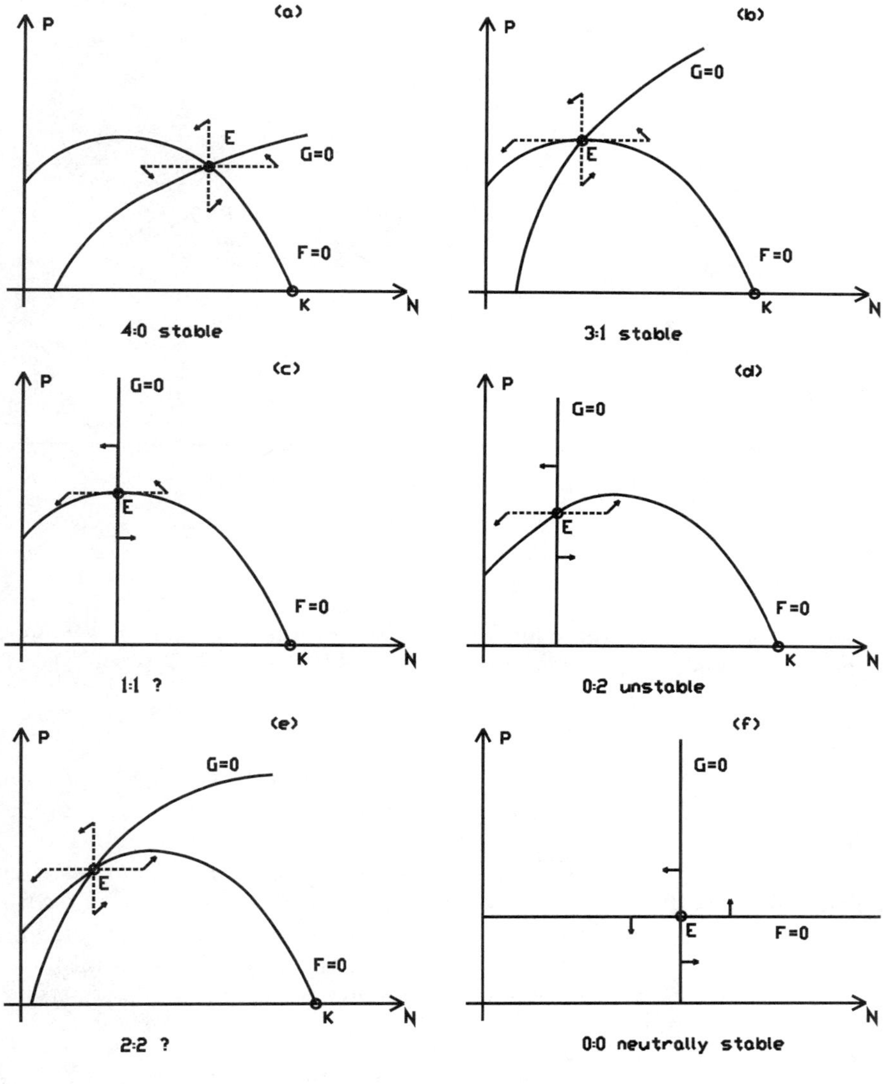

Figure 2.2.3: The equilibrium of the predator-prey system (2.2.1) and the "method of the little arrows". The scores are those of a match between stability (meaning asymptotic stability) and instability. Stability scores one if an arrow shows toward E and instability scores one if an arrow shows away from E. If the arrow is perpendicular to the line through E then neither stability nor instability scores. (c) and (d) belong to systems, where G does not depend on P; (f) belongs to the simplest Lotka-Volterra system (2.2.3) where E is Lyapunov stable but not asymptotically stable.

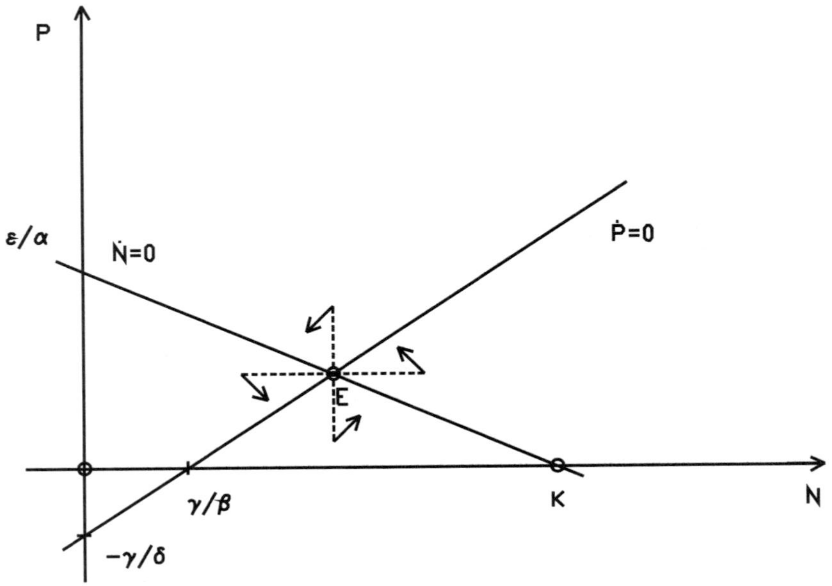

Figure 2.2.4: The interior equilibrium point of the general Lotka-Volterra predator-prey system (2.2.1) with the "method of the little arrows" showing its asymptotic stability.

to decrease the distance, that is, tries to drive the system back into equilibrium, then we may expect that the equilibrium is asymptotically stable. This method is shown in Fig. 2.2.3.

Although, the "method of little arrows" does not yield a proof of stability or instability and may well be misleading, nevertheless it provides one with a guess and then one has "but" prove the guess according to Polya's maxim to mathematicians: "Guess and prove!" To find the stability condition for the equilibrium point E we determine the Jacobi matrix of system (2.2.4)

$$J = \begin{bmatrix} F(N,P) + NF'_N(N,P) & NF'_P(N,P) \\ PG'_N(N,P) & G(N,P) + PG'_P(N,P) \end{bmatrix},$$

and the characteristic equation at $E = (\overline{N}, \overline{P})$,

$$\lambda^2 - \left(\overline{N}F'_N(\overline{N},\overline{P}) + \overline{P}G'_P(\overline{N},\overline{P})\right)\lambda + \overline{N}\overline{P}\left(F'_N(\overline{N},\overline{P})\,G'_P(\overline{N},\overline{P})\right.$$

$$\left. - F'_P(\overline{N},\overline{P})\,G'_N(\overline{N},\overline{P})\right) = 0\,. \tag{2.2.5}$$

By the assumptions imposed upon functions F and G, if F'_N is negative at E then both the constant term and the coefficient of λ are positive and by Theorem A1.1.2 E is asymptotically stable. This justifies Fig. 2.2.3a. The quantity F'_N at E is negative iff the equilibrium point is on the descending branch of the prey curve, because F is increasing going into the interior of the domain bounded by the prey curve, and this implies that its gradient $[F'_N, F'_P]$ points into the interior of this domain. What we achieve is called the *Rosenzweig-MacArthur graphical criterion of stability*–the equilibrium of a predator-prey system is asymptotically stable if it lies on the descending branch of the prey curve (Rosenzweig-MacArthur 1963; Freedman 1976). If E is at the maximum point of the prey curve and G depends on P, then though F'_N is zero, G'_P is negative so that again both the constant term and the coefficient of λ are positive. This justifies Fig. 2.2.3b. If in this case G does not depend on P, then the coefficient of λ is zero, so that the characteristic polynomial has a pair of pure imaginary roots and thus the stability problem cannot be settled this way in accordance to Fig. 2.2.3c. One may similarly justify Fig. 2.2.3d,e.

If for the stability problem of the equilibrium point $(K, 0)$ representing the absence of a predator, we substitute this point into the Jacobi matrix, the characteristic equation becomes

$$\lambda^2 - \lambda (G(K,0) + KF'_N(K,0)) + KF'_N(K,0)\, G(K,0) = 0\,.$$

The function $F(N, 0)$ is decreasing at $N = K$, so that normally $F'_N(K, 0)$ is negative. If the predator curve meets axis N left of point K then, as this was already mentioned, the interior equilibrium point $E = (\overline{N}, \overline{P})$ exists and $G(K, 0)$ is positive. This means that the constant term in the characteristic equation is negative and the equilibrium $(K, 0)$ is unstable (actually, it is a saddle point). If the predator curve meets axis N right of point K then there is no interior equilibrium and $G(K, 0)$ is negative; hence, all coefficients in the characteristic polynomial are positive, the equilibrium $(K, 0)$ is asymptotically stable, and the predator dies out. This is what one would expect, because in this case the predator may start to increase its quantity only when the quantity of prey is above K but the prey quantity can stay above the carrying capacity only temporarily.

It is worthwhile to return to Fig. 2.2.3c,d and explore what happens in this case (when G does not depend on P) if the system depends on a parameter (K, say) and as this parameter is varied continuously the equilibrium point E is wandering on the prey curve and having reached the maximum point moves from the descending branch of the curve to the ascending branch and loses its stability as a consequence. If we look at the characteristic polynomial (2.2.5) where G'_P is now zero we see that the constant term always remains positive while the coefficient of λ changes its sign at the maximum point. The eigenvalues are continuous functions of the coefficients of the characteristic polynomial, which are continuous functions of the parameter, so that if the real parts of the eigenvalues change their sign from negative to positive and the constant term remains positive so that no eigenvalue assumes the value zero in the process, this may happen only one way–at a certain value of the parameter (when the equilibrium is passing through the maximum point of the prey curve) both eigenvalues cross the imaginary axis away from the origin. At the *critical parameter value* (when the real parts are zero), the linearized system has a pair of conjugate imaginary eigenvalues, that is, every solution of the system linearized at the equilibrium point is periodic. This is a typical situation for an *Andronov-Hopf bifurcation* (see Appendix 2.3); the nonlinear system (2.2.4) will have some small amplitude periodic solution around the equilibrium point. If one can establish that the bifurcation is *supercritical* then although the equilibrium is destabilized (it is already on the ascending branch of the prey curve near to the maximum point) the quantities of predator and prey will periodically oscillate in a stable way.

The same question may be raised concerning Fig. 2.2.3a,b,e when G depends on P. It is clear from looking at the signs of the coefficients of the characteristic polynomial that equilibrium E will now stay asymptotically stable even if it moves slightly left of the maximum point but if it moves farther to the left it may lose its stability. The trouble now is that one cannot tell whether the constant term or the coefficient of λ will be zero first, as this depends on the particular system considered.

2.3 Two Interacting Species–Competition and Cooperation

If two species are chosen that compete for the same resources and the effect of other species is neglected, we speak of a two dimensional (2D) competitive system. Competition means that the increase of the numbers of the first species has an adverse effect on the growth rate of the second one and vice versa. If the resources for which the species are competing are not specified and competition is expressed by this adverse effect then the situation can be modeled by a 2D system of differential equations. Following logistic dynamics, we assume first that the per capita growth rates of the competing species are linear functions of the quantities and that the two species, if left alone, can survive. This way we arrive at the *competitive Lotka-Volterra system*

$$\dot{N}_1 = N_1\left(r_1 - a_{11}N_1 - a_{12}N_2\right), \qquad \dot{N}_2 = N_2\left(r_2 - a_{21}N_1 - a_{22}N_2\right) \quad (2.3.1)$$

where $N_1(t)$ and $N_2(t)$ are the quantities of the two species at time t, $r_1 > 0$ and $r_2 > 0$ are the intrinsic growth rates of the respective species, the entries of the matrix $A = [a_{ik}]$ are positive, a_{11} and a_{22} represent the strength of the *intraspecific competition* (the competition within the species, r_1/a_{11}, r_2/a_{22} are the carrying capacities of the respective species), and a_{12} and a_{21} represent the strength of the *interspecific competition* (the competition between the two species, for instance, a_{12} is the amount by which one unit of species 2 decreases the per capita growth rate of species 1).

Possible equilibria of the system and their stability are explored here. We will condider only generic cases and assume that $\det A \neq 0$. There are four equilibria, $E_0 = (0,0)$, $E_1 = (r_1/a_{11}, 0)$ representing the absence of species 2,

$E_2 = (0, r_2/a_{22})$ representing the absence of species 1, and

$$E = (\overline{N}_1, \overline{N}_2) = \left(\frac{r_1 a_{22} - r_2 a_{12}}{\det A}, \frac{r_2 a_{11} - r_1 a_{21}}{\det A} \right),$$

where the latter one is in the interior of the positive quadrant of the N, P plane iff either

$$a_{12}/a_{22} < r_1/r_2 < a_{11}/a_{21}\,, \tag{A}$$

or

$$a_{11}/a_{21} < r_1/r_2 < a_{12}/a_{22}\,. \tag{B}$$

In case neither (A) nor (B) is fulfilled, for example,

$$r_1/r_2 < a_{12}/a_{22} \quad \text{and} \quad r_1/r_2 < a_{11}/a_{21}\,, \tag{C}$$

then there is no equilibrium with positive coordinates representing the coexistence of the two species. (A completely symmetric case (C) is obtained by interchanging the indices in the previous inequalities.)

To determine the conditions of stability of these equilibria we deduce the Jacobian matrix of the system,

$$J(N_1, N_2) = \begin{bmatrix} r_1 - 2a_{11}N_1 - a_{12}N_2 & -a_{12}N_1 \\ -a_{21}N_2 & r_2 - a_{21}N_1 - 2a_{22}N_2 \end{bmatrix},$$

substitute the coordinates of the respective equilibria and determine the eigenvalues. In all cases the origin E_0 is unstable (as a matter of fact, it is an unstable node). We write out only the characteristic polynomial at E,

$$p(\lambda) = \lambda^2 + \lambda(a_{11}\overline{N}_1 + a_{22}\overline{N}_2) + \overline{N}_1\overline{N}_2 \det A\,.$$

For (A), equilibria E_1 and E_2 are unstable, and it is easy to see that (A) implies that $\det A > 0$. This means by Theorem A1.1.2 that E is asymptotically stable, that is, the two competing species may coexist in the long run. In case (B) both E_1 and E_2 are asymptotically stable and $\det A < 0$, that is, E is unstable (as a matter of fact, it is a saddle point). This means that in this case one of the species dies out, which one depends on the initial conditions–there is a *separatrix* going through point E, from above this curve all trajectories go to E_2, from below they go to E_1. In this case the system is said to be *bistable.* Note that the condition of stability of E, $\det A > 0$ is equivalent to

$$a_{11}a_{22} > a_{12}a_{21}\,. \tag{2.3.2}$$

This last condition means that the overall intraspecific competition is stronger than the interspecific competition. Thus, *if the intraspecific competition is stronger than the interspecific competition then the two species may coexist in a stable way; in the opposite case one of them dies out.* In case (C) there is no interior equilibrium point, the zero isoclines of the two species do not intersect in the interior of the positive quadrant. In this case species 2 outcompetes species

1; no matter from where we start in the interior of the positive quadrant all trajectories tend to E_2. This is fairly intuitive, as the inequalities characterizing case (C) can be put in the form

$$r_1 a_{22} < r_2 a_{12} , \qquad r_1 a_{21} < r_2 a_{11} ,$$

in which the first one means that the growth rate of species 1 and the intraspecific competition in species 2 are relatively small compared to the growth rate of species 2 and the adverse effect that species 2 has on species 1; the second one means that the growth rate of species 1 and the adverse effect species 1 has on species 2 are relatively small compared to the growth rate of species 2 and the intraspecific competition in species 1. Both inequalities clearly express the advantage of species 2 over species 1. See Fig. 2.3.1.

To be sure, a stability analysis based on the eigenvalues of the Jacobian at different equilibria yields only local results. In order to establish such statements that in case (A) all trajectories from the interior of the positive quadrant tend to E, or that in case (C) all trajectories tend to E_2, say, one has to show that all the solutions of system of equations (2.3.1) are bounded on $(0, \infty)$ and the system has no closed trajectories in the positive quadrant. The first statement is clear if one looks at the system, because both N_1 and N_2 are decreasing if (N_1, N_2) is "northeast" from the straight lines $a_{11}N_1 + a_{12}N_2 = r_1$ and $a_{21}N_1 + a_{22}N_2 = r_2$. The statement that system (2.3.1) has no closed trajectory follows from the Poincaré Bendixson theory (see Appendix 2.2) for case (C) when there is no equilibrium in the positive quadrant and from an inspection of the phase portrait in cases (A) and (B).

The situation is different if the competition goes on for a single resource without which none of the competitors can survive. If the competitors' per capita growth rates depend linearly on the quantity of an available resource and this resource can be depleted then the fitter of the two competitors will survive and the other one will die out. This law is called the *competitive exclusion principle.* We will show this phenomenon in the sequel. The quantities of the competitors at time t will be denoted by $P_1(t), P_2(t)$, their death rates (mortalities) by $d_1, d_2 > 0$, and the quantity of the resource is given by a function $s(P_1, P_2)$ defined in the closed positive quadrant of the P_1, P_2 plane. This function is supposed to be continuously differentiable, positive in a bounded domain containing the origin, and negative outside this domain (negative outside a circle with the center in the origin, for example). Then the system describing the dynamics is

$$\dot{P}_1 = P_1 (-d_1 + b_1 \ s(P_1, P_2)) , \quad \dot{P}_2 = P_2 (-d_2 + b_2 \ s(P_1, P_2)) , \qquad (2.3.3)$$

where the quantities $b_1, b_2 > 0$ are the efficiencies by which the two competitors make use of the resource. We assume that in a neighborhood of the origin the right-hand sides are positive, that is, the competitors may survive at low densities. It is clear that every solution is bounded in the future, because the right-hand sides are negative far from the origin. Now, let $c_1 > 0$ and $c_2 > 0$ such that $b_1 c_1 - b_2 c_2 = 0$, divide the first and the second equation by P_1 and

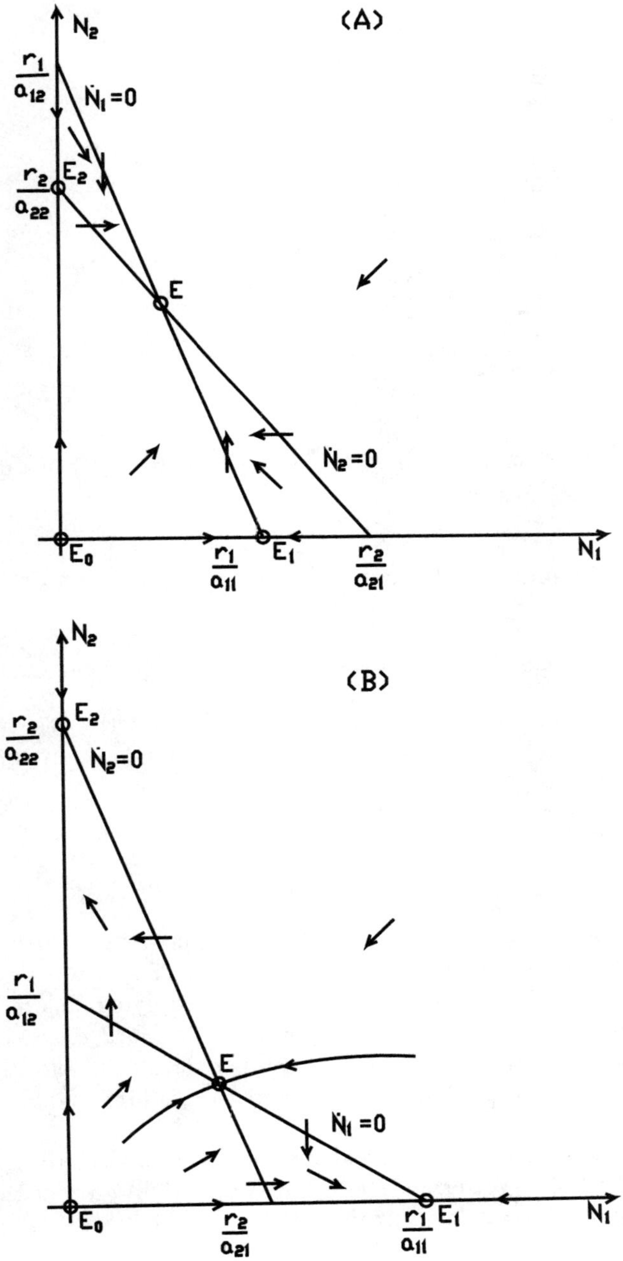
(A)
N2
r1/a12
N1=0
r2/a22
E2
E
N2=0
E0
r1/a11
E1
r2/a21
N1
N2
(B)
E2
r2/a22
N2=0
r1/a12
E
N1=0
E0
r2/a21
r1/a11
E1
N1

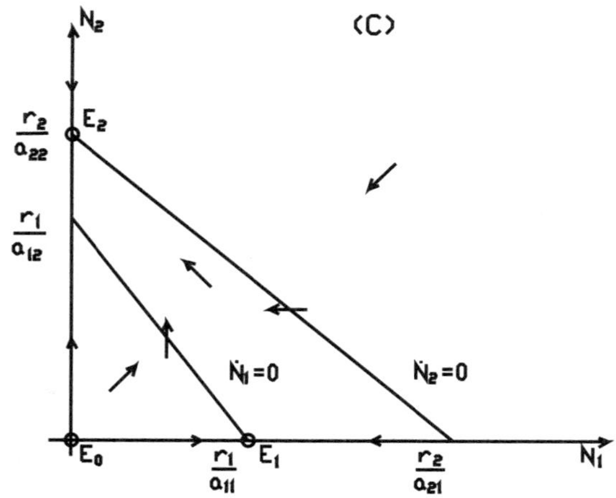

Figure 2.3.1: Two competitors: (A) stable coexistence; (B) bistability; and (C) species 2 outcompetes species 1.

P_2, respectively, multiply the first equation by c_1, the second by $-c_2$, and add them. W e obtain

$$c_1 \dot{P}_1/P_1 - c_2 \dot{P}_2/P_2 = -(d_1 c_1 - d_2 c_2) = -a \ ,$$

where we may assume that $a > 0$ (if this were not the case we use $-c_1, c_2$ instead of $c_1, -c_2$; the nongeneric case $a = 0$ is excluded). Integrating the last equation we get

$$\ln \left(P_1^{c_1}(t)/P_2^{c_2}(t)\right) = -at + c$$

or

$$P_1^{c_1}(t)/P_2^{c_2}(t) = Ce^{-at} \to 0 \quad \text{as} \quad t \to \infty \, .$$

Because the solutions are bounded this implies that P_1 tends to zero as t tends to infinity, that is, the first competitor dies out. This result is intuitively reasonable. By our assumptions $d_1 c_1 > d_2 c_2$. Multiplying this inequality by b_1 and using the fact that $b_1 c_1 = b_2 c_2$ we get $b_2 c_2 d_1 > b_1 d_2 c_2$, or $b_1/d_1 < b_2/d_2$, meaning that the efficiency of the first competitor is lo w and its death rate is high while the efficiency of the second competitor is high and the death rate of the latter is low.

Note that when function s depends linearly on P_1 and P_2 then system (2.3.3) reduces to the Lotka-Volterra system (2.3.1) case (C) (without the assumption $r_1, r_2 > 0$) and the zero isoclines are parallel straight lines, hence, they do not intersect at all.

The competitive exclusion principle also applies under similar linearity conditions in higher dimensional cases when there are more competitors than resources (see McGehee-Armstrong, 1977; Hofbauer-Sigm und, 1988). It is not necessarily true if the dependence is not linear and the resources are regenerative.

We say that the two species are *cooperative (mutualists, symbiotic)* if the presence and the increase of the first one is beneficial to the growth rate of the second and vice versa. Thus, the *two-dimensional cooperative Lotka-Volterra system* has the form

$$\dot{N}_1 = N_1 (r_1 - a_{11}N_1 - a_{12}N_2) , \quad \dot{N}_2 = N_2 (r_2 - a_{21}N_1 - a_{22}N_2) \tag{2.3.4}$$

where it is assumed that $a_{12}, a_{21} < 0$ and $a_{11}, a_{22} \geq 0$, which expresses the fact that the system is cooperative but there may be intraspecific competition within each species.

The first case to be considered is when each species survives if left alone and follows the logistic dynamics, that is, the intrinsic growth rates are positive, $r_1, r_2 > 0$. This is sometimes called *facultative mutualism*. It is easy to see that in this case the origin is a repeller and the eigenvalues of the system linearized at the origin are $r_1 > 0$ and $r_2 > 0$. The equations of the $\dot{N}_1$ and $\dot{N}_2$ zero isoclines are (apart from the coordinate axes)

$$N_2 = -N_1 a_{11}/a_{12} + r_1/a_{12} \quad \text{and} \quad N_2 = -N_1 a_{21}/a_{22} + r_2/a_{22} , \tag{2.3.5}$$

respectively, provided that $a_{22} > 0$. By our assumptions the first line crosses the axis N_2 below the origin and the second above it and both slopes are positive, so that the two lines have a point of intersection in the positive quadrant if and only if the slope of the first line is larger than that of the second one $-a_{11}/a_{12} > -a_{21}/a_{22}$, or

$$\det A = \det[a_{ik}] = a_{11}a_{22} - a_{12}a_{21} > 0 . \tag{2.3.6}$$

Intuitively, this condition means that the overall intraspecific competition within the species $a_{11}a_{22}$ is stronger than the level of mutualism $a_{12}a_{21}$. The point of intersection, the single equilibrium in the positive quadrant is

$$(\overline{N}_1, \overline{N}_2) = \left(\frac{r_1 a_{22} - r_2 a_{12}}{\det A}, \frac{r_2 a_{11} - r_1 a_{21}}{\det A} \right) .$$

Looking at the phase portrait of the system (Fig. 2.3.2 (A)) we see that all trajectories from the interior of the positive quadrant eventually enter the acute angle sections bounded by the two zero isoclines and then tend towards this equilibrium point, so that $(\overline{N}_1, \overline{N}_2)$ is globally asymptotically stable in the open positive quadrant. If the inequality in (2.3.6) is reversed then the two isoclines do not intersect in the positive quadrant and all trajectories from the interior of the positive quadrant enter the widening strip bounded by these lines and then tend to infinity (see Fig. 2.3.2 (B)).

The situation is different if the cooperation is not facultative, that is, the species die out without the mutualist. If this is the case we speak about *obligatory cooperation*. System (2.3.4) is considered again but now r_1 and r_2 are supposed to be negative. This means that now the first line of (2.3.5) crosses

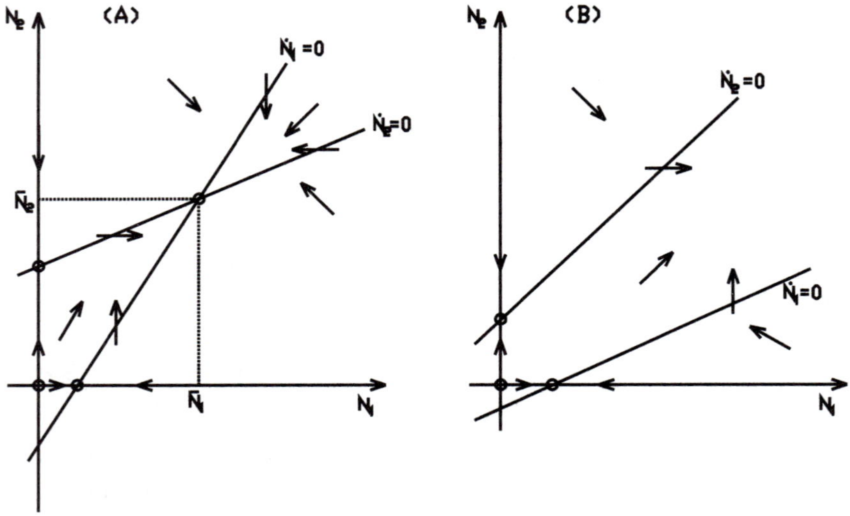

Figure 2.3.2: Facultatively cooperating species by system (2.3.4) $(r_1, r_2 > 0)$. (A) the intraspecific competition is stronger than the cooperation; and (B) the cooperation is stronger than the intraspecific competition.

the axis N_2 above the origin and the second one below it, so that the condition of having a point of intersection in the positive quadrant is

$$\det A = \det[a_{ik}] = a_{11}a_{22} - a_{12}a_{21} < 0\,, \tag{2.3.7}$$

and, indeed, in this case the coordinates of the equilibrium $(\overline{N}_1, \overline{N}_2)$ are positive. If we look at the phase portrait (Fig. 2.3.3 (A)) it is clear that the origin is now asymptotically stable (the eigenvalues r_1 and r_2 are negative), and that the equilibrium $(\overline{N}_1, \overline{N}_2)$ is a saddle point (this can also be seen if one writes out the Jacobian matrix at $(\overline{N}_1, \overline{N}_2)$ and realizes that the product of the two eigenvalues is the expression (2.3.7)). The phase portrait shows that all trajectories starting in the interior of the positive quadrant eventually enter the acute angle sections bounded by the two zero isoclines. Further, if they start below the ingoing trajectories of the saddle, that is, the initial quantities of the two species are low or at least one of them is very low, then both species die out; however, if the initial quantities are sufficiently large then both species grow unlimited. Note that, as previously mentioned, (2.3.7) shows that cooperation is now stronger than the intraspecific competition within the species.

If inequality (2.3.7) is reversed, meaning the intraspecific competition is stronger than the cooperation, then the two isoclines do not intersect in the positive quadrant and the phase portrait is as shown Fig. 2.3.3(B), where all trajectories enter the widening strip between the two isoclines and then tend to the origin. This means that in spite of cooperation the two species die out.

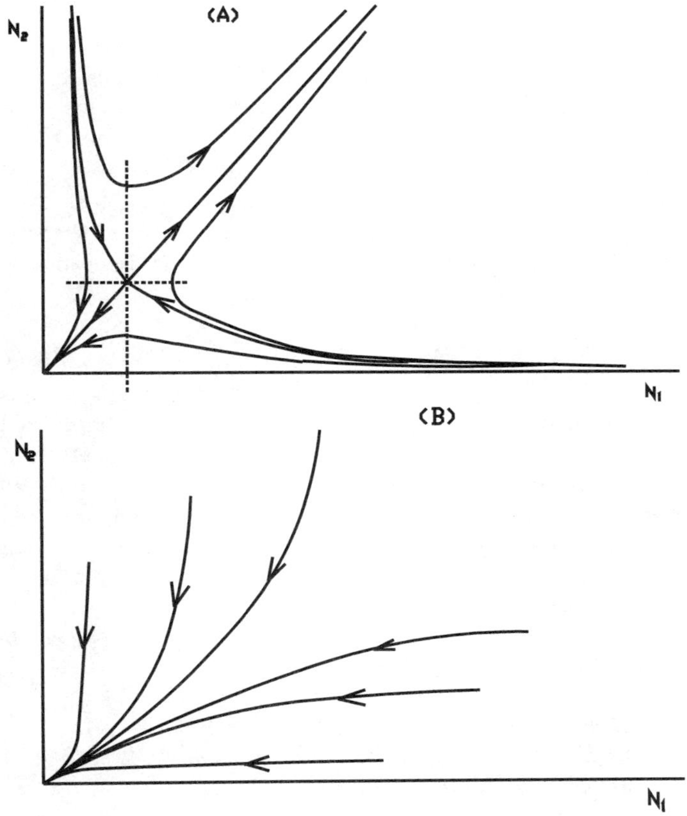

Figure 2.3.3: Obligatory cooperation by system (2.3.4) ($r_1, r_2 < 0$). (A) cooperation is stronger than intraspecific competition $a_{11} = a_{22} = 0$, $a_{12} = -1$, $a_{21} = -2$, $r_1 = r_2 = -1$; and (B) intraspecific competition is stronger than cooperation $a_{11} = a_{22} = 2$, $a_{12} = -1$, $a_{21} = -2$, $r_1 = r_2 = -1$ (PHASER).

Facultative cooperation may exist between a plant and a certain insect (e.g., bees), where pollination is the result of the cooperation, and obligatory cooperation exists between a mammal and intestinal bacteria. According to May (1981), obligatory mutualism is more prevalent in tropical regions.

The Lotka-Volterra models treated in this Section may be considered fairly unrealistic allowing for the possibility in some cases that species could grow unbounded. This disadvantage can be overcome by introducing per capita growth rates that depend nonlinearly on the quantities of the species. We shall return to this problem when treating higher dimensional systems in the next Section.

2.4 More Species

In this Section ecological systems with more than two species living together will be considered . Because of lack of space it is impossible to give a systematic treatment of all the mathematical tools that may be applied but we shall try to survey this vast field by using carefully chosen examples.

Suppose that n different species live together in a habitat and that the growth rate of each depends on the quantity of every one of them. Denoting the quantity of the ith species by x_i, $(i = 1, 2, \ldots, n)$, the dynamics is governed by the so-called *Kolmogorov system of differential equations*

$$\dot{x}_i = x_i f_i (x_1, x_2, \ldots, x_n) \ , \qquad i = 1, 2, \ldots, n \, . \tag{2.4.1}$$

The main feature of this autonomous system of differential equations is that the coordinate hyperplanes $x_i = 0$, $(i = 1, 2 \ldots, n)$ and the orthants of the space bounded by them are invariant, that is, if an initial condition is given in the interior of the positive orthant $\mathbb{R}^n_+ = \{(x_1, x_2, \ldots, x_n) \in \mathbb{R}^n : x_1 \geq 0, x_2 \geq 0, \ldots, x_n \geq 0\}$, then the corresponding trajectory stays in this orthant for the entire future (and past). The character of the relationship between species j and species k is determined by the response of the per capita growth rate $\dot{x}_j/x_j$ to the increase of x_k and vice versa by interchanging the indices j and k, that is, the relationship depends on the sign of the derivative of f_j with respect to x_k, and vice versa. There are three important special cases:

(i) If for each $j \neq k$ and in the whole positive orthant $f'_{jx_k} f'_{kx_j} \leq 0$, then we say that (2.4.1) represents a *predator-prey system*; if $f'_{jx_k} < 0$ and $f'_{kx_j} > 0$ then species j is prey for species k.

(ii) If for each $j \neq k$ and in the whole positive orthant $f'_{jx_k} < 0$, then we say that (2.4.1) represents a *competitive system.*

(iii) If for each $j \neq k$ and in the whole positive orthant $f'_{jx_k} > 0$, then we say that (2.4.1) represents a *cooperative system.*

It may happen that $f'_{1x_2} < 0$, $f'_{2x_1} > 0$, $f'_{2x_3} < 0$, $f'_{3x_2} > 0$, $f'_{3x_4} < 0$, $f'_{4x_3} > 0$ and species 4 is no prey for any other species. In this case we say that species 1,2,3, and 4 form a *food chain* where species 4 is the *top predator.* Naturally it may happen that the system of equations (2.4.1) does not belong to any of the three forementioned types containing predator-prey relations between some species and competition and cooperation between some other ones. A fairly advanced theory has been worked out for competitive and cooperative systems by M. Hirsch , Hal L. Smith and others (see Smith H.L. 1995). Because of lack of space we cannot go into this here. Predator-prey systems have no such general theory.

Where the system is Lotka-Volterra, that is,

$$\dot{x}_j = x_j \left(r_j - \sum_{1}^{n} a_{jk} x_k \right) \ , \quad j = 1, 2, \ldots, n \, , \tag{2.4.2}$$

the characterization of the three types boils down to the signs of the entries of the *interaction* (or *community*) *matrix* $A = [a_{jk}]$. System (2.4.2) is *predator-prey, competitive, cooperative,* respectively, if for all $j \neq k$: (i) $a_{jk}a_{kj} \leq 0$; (ii) $a_{jk} > 0$; and (iii) $a_{jk} < 0$.

Rather than discuss of these general systems, we shall present some basic ecological phenomena through concrete models. A three-species model will be studied first, and the competition of an "*r-strategist*" and a "*K-strategist*" predator for a single regenerating prey species will be described. Somewhat simplifying the concepts, an *r-strategist* is a species that tries to ensure its survival by having a relatively high growth rate and a *K-strategist* is a species that consumes less, has a lower growth rate and is able to raise its offspring on a scarce supply of food. For more accurate concepts see May (1981). The quantities of prey and the two predators at time t will be denoted by $S(t), x_1(t), x_2(t)$, respectively, the intrinsic growth rate of the prey and the carrying capacity will be denoted by γ and K, respectively, and the maximal birth rate and the (constant) death rate of predator i will be m_i and d_i, respectively. All these constants are supposed to be positive. The system considered is

$$\begin{aligned} \dot{S} &= \gamma S(1 - S/K) - m_1 x_1 S/(a_1 + S) - m_2 x_2 S/(a_2 + S) \\ \dot{x}_1 &= m_1 x_1 S/(a_1 + S) - d_1 x_1 \\ \dot{x}_2 &= m_2 x_2 S/(a_2 + S) - d_2 x_2 , \end{aligned} \tag{2.4.3}$$

where $a_i > 0$ is the *half saturation constant* of predator i. This means that when the quantity of prey S reaches value a_i then the per capita birth rate of predator i reaches half of the maximal birth rate, as can be seen if $S = a_i$ ($i = 1$ and 2, respectively) is substituted into the first term of the second, and third equations of the system, respectively. Clearly, the less a_i is the less food is needed for predator i, so that a predator with a small half saturation constant can be considered a K-strategist. The first terms in the second and third equations are called *Holling II type functional responses.* They should be multiplied by some positive constants < 1 to allow for losses when biomass from the prey is built into the predator species but these "loss factors" could be transformed out of the equations. The Holling type functional response is more realistic than the $m_i x_i S$ that occurs in a Lotka-Volterra system because the latter tends to infinity as prey quantity tends to infinity. If one takes into consideration that no matter how much food is available the predator cannot handle more than a certain limited quantity in unit time, then the Holling type terms are satisfactory because they increase with S but do not tend to infinity and are concave down, thereby representing a saturation effect when there is too much food. System (2.4.3) has been dealt with by several authors (see Farkas, 1984b, and the references therein). The growth rate of predator i is zero when the right-hand side of the corresponding equation is zero, that is,

$$m_i x_i S/(a_i + S) - d_i x_i = 0 \quad \text{or} \quad S = \lambda_i := a_i d_i/(m_i - d_i) \ .$$

If S is above this value the growth rate is positive and if it is below it the predator is going to die out. We are to consider here only the nongeneric case in which these threshhold values for the two predators are equal,

$$\lambda := \lambda_1 = a_1 d_1/(m_1 - d_1) = \lambda_2 = a_2 d_2/(m_2 - d_2) . \tag{2.4.4}$$

If we look at the first equation in the system (2.4.3) we see that without predators the prey quantity follows the logistic dynamics, that is, it cannnot stay above K in the long run. With predation its value is still less, so that if $\lambda \geq K$ both predators die out. It is also clear that if the maximal birth rate of the predator is less than its death rate the predator dies out (in this case the right-hand side of the corresponding predator equation is always negative). Thus, in order to have a system in which survival of the species is possible we assume that

$$m_i > d_i \quad (i = 1, 2) \quad \text{and} \quad 0 < \lambda < K . \tag{2.4.5}$$

Introducing the parameter $\beta_i = m_i - d_i$ under conditions (2.4.4) and (2.4.5) our system takes the form

$$\begin{aligned} \dot{S} &= \gamma S\left(1 - S/K\right) - m_1 x_1 S/\left(a_1 + S\right) - m_2 x_2 S/\left(a_2 + S\right) \\ \dot{x}_1 &= \beta_1 x_1 \left(S - \lambda\right)/\left(a_1 + S\right) \\ \dot{x}_2 &= \beta_2 x_2 \left(S - \lambda\right)/\left(a_2 + S\right) \end{aligned} \tag{2.4.6}$$

where because of our assumptions $\beta_i > 0$ $(i = 1, 2)$. We assume that $a_1 > a_2$. This implies that $m_1/d_1 > m_2/d_2$ because by (2.4.4)

$$\lambda = a_1/(m_1/d_1 - 1) = a_2/(m_2/d_2 - 1) .$$

Thus, the birth rate relative to the death rate for predator 1 is higher than for predator 2 while the half saturation constant of predator 2 is lower than that of predator 1. This means that we may identify predator 1 as an r-strategist and predator 2 as a K-strategist. The equilibria of system (2.4.6) are $(0, 0, 0)$ and $(K, 0, 0)$ and the points of the straight line segment

$$L = \left\{ (S, x_1, x_2) \in \mathbb{R}^3 : S = \lambda,\ x_1 \geq 0,\ x_2 \geq 0,\ \frac{m_1 x_1}{a_1 + \lambda} + \frac{m_2 x_2}{a_2 + \lambda} = \gamma\left(1 - \frac{\lambda}{K}\right)\right\} .$$

We shall now study the stability of the equilibria. By linearization it is easy to see that the origin is always unstable and that $(K, 0, 0)$ is unstable if the second inequality of (2.4.5) holds (if it is reversed, i.e., $\lambda > K$, then it is asymptotically stable, which as previously mentioned, implies that the predators are dying out). We choose now an arbitrary point on the line segment $(\lambda, \xi_1, \xi_2) \in L$, linearize the system at this point, and determine the characteristic polynomial. By performing the calculations we obtain

$$D(\mu) = \mu\left(\mu^2 + \mu\lambda\left(\frac{\gamma}{K} - \frac{m_1 \xi_1}{(a_1 + \lambda)^2} - \frac{m_2 \xi_2}{(a_2 + \lambda)^2}\right)\right.$$

$$+\lambda\left(\frac{\beta_1 m_1\xi_1}{(a_1+\lambda)^2}+\frac{\beta_2 m_2\xi_2}{(a_2+\lambda)^2}\right)\bigg)\ .$$

The quadratic polynomial in brackets is stable iff

$$m_1\xi_1/(a_1+\lambda)^2 + m_2\xi_2/(a_2+\lambda)^2 < \gamma/K\ . \tag{2.4.7}$$

Now, if $\lambda < K < a_2 + 2\lambda$ then

$$\begin{aligned} &m_1\xi_1/\left(a_1+\lambda\right)^2 + m_2\xi_2/\left(a_2+\lambda\right)^2 \\ &\leq \left(m_1\xi_1/\left(a_1+\lambda\right) + m_2\xi_2/\left(a_2+\lambda\right)\right)/\left(a_2+\lambda\right) \\ &= \gamma\left(1-\lambda/K\right)/\left(a_2+\lambda\right) < \gamma/K\ . \end{aligned}$$

By an analogous estimate, if $K > a_1 + 2\lambda$ then

$$m_1\xi_1/(a_1+\lambda)^2 + m_2\xi_2/(a_2+\lambda)^2 > \gamma/K\ .$$

Thus, for low carrying capacity, $\lambda < K < a_2 + 2\lambda$, each interior equilibrium on the line segment L has a zero eigenvalue and two eigenvalues with negative real parts. One may prove that each equilibrium point on L is stable in the Lyapunov sense, *the segment L is an attractor of the system*, and every trajectory that starts in the interior of the positive octant tends to some point on L. This means that if food is scarce both the r- and K-strategists may live together in the long run in a steady state that depends on the initial values of the species. It is to be noted that Lyapunov stability (without asymptotic stability) cannot be relied upon if, say, small stochastic variations are to be taken into account; the equilibrium on L may wander and eventually may reach one of the end points of the segment, which means the disappearance of one of the predator species. For high carrying capacity, $K > a_1 + 2\lambda$, all the equilibria on the segment are unstable.

We turn now to the case where $a_2 + 2\lambda \leq K \leq a_1 + 2\lambda$. It is easy to see that as the point (λ, ξ_1, ξ_2) moves along the segment from left to right from the point $(\lambda, 0, \gamma(a_2+\lambda)(K-\lambda)/(m_2K))$ to $(\lambda, \gamma(a_1+\lambda)(K-\lambda)/(m_1K), 0)$ the expression on the left-hand side of (2.4.7) is decreasing. If K is fixed in the interval $(a_2+2\lambda, a_1+2\lambda)$ then there is a point $B(K) = (\lambda, \xi_1(K), \xi_2(K))$ on L at which (2.4.7) turns into an equation. The points of L to the left of $B(K)$ are unstable, those to the right of $B(K)$ are stable, and that part of L which is to the right of point $B(K)$ is still an attractor of the system (see Fig. 2.4.1). If we move K in the interval $(a_2+2\lambda, a_1+2\lambda)$ from left to right the point $B(K)$ moves along L from the left end to the right, and the points left behind are destabilized. We call this phenomenon a *zip bifurcation.* (In the process the straight line L itself is also displaced in a parallel way but this is irrelevant.) From the point of view of the competition, as the quantity of available food is increasing the K-strategist is losing ground and those equilibria where the relative quantity of K-strategist is high compared to the quantity of r-strategist are the first to be destabilized. When K reaches value $a_1 + 2\lambda$ all interior equilibria have already been destabilized and the only stable equilibrium remaining is the endpoint of

L in the S, x_1 plane. This means that at this carrying capacity value the K-strategist dies out. One may prove that if K is increased further then even the equilibrium in the S, x_1 plane gets destabilized but the prey and the r-strategist continue to coexist in a periodic way (an Andronov-Hopf bifurcation occurs, see Appendix 2.3).

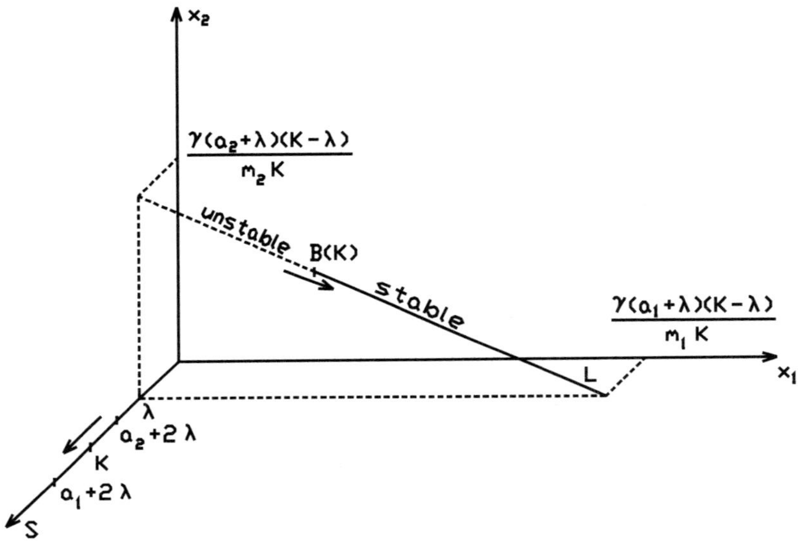

Figure 2.4.1: Zip bifurcation in system (2.4.6). As K moves forward in the interval $(a_2 + 2\lambda, a_1 + 2\lambda)$ the point $B(k)$ moves along L from left to right and leaves behind destabilized equilibria.

Note that the same phenomenon occurs in a much wider class of models (see Farkas 1987; 1994).

We turn now to a more general model of competition of t wo predator species for a single regenerating prey. The per capita growth rates will be assumed as arbitrary functions that satisfy certain basic requirements and intraspecific competition will be assumed in the predators, which was not the case in the previous model. (Because of the latter assumption the previous model is not a special case of the one that follows.) Our main purpose is to show how the Rosenzweig-MacArthur graphical criterion (see Section 2.2) generalizes to situations in which more than two species are involved. We denote now the quantity of prey at time t by $x(t)$ and that of the two predators by $y_1(t)$ and $y_2(t)$. The system to be studied is

$$\dot{x} = xF(x, y_1, y_2, K), \quad \dot{y}_1 = y_1 G_1(x, y_1), \quad \dot{y}_2 = y_2 G_2(x, y_2) \qquad (2.4.8)$$

where the following assumptions are made:

$$(x - K)F(x, 0, 0, K) < 0, \quad x \geq 0, \quad x \neq K > 0, \qquad (2.4.9)$$

(in the absence of predation, prey grows until it reaches the carrying capacity),

$$F'_{y_i}(x, y_1, y_2, K) < 0\,, \qquad (i = 1, 2)\,, \tag{2.4.10}$$

(the y-s are predators of x),

$$\begin{aligned} G_i(0, y_i) < 0, \quad G'_{iy_i}(x, y_i) \le 0, \\ G'_{1y_1} + G'_{2y_2} < 0, \quad G'_{ix}(x, y_i) > 0\,, \quad (i = 1, 2)\,, \end{aligned} \tag{2.4.11}$$

(the meaning of the last four inequalities is: predators cannot survive without prey; there might be intraspecific competition within the predator species and there is one indeed in one of the species at least; and the increase of prey is increasing the growth rate of the predators). It is to be noted that the two predator species compete only by consuming the same prey (y_1 does not occur explicitly in the equation for y_2 and vice versa). We assume further that the two predator isoclines $G_i(x, y_i) = 0$, $(i = 1, 2)$ (cylinder surfaces) intersect in a curve in the positive octant of x, y_1, y_2 space and that this curve intersects the prey zero isocline surface $F(x, y_1, y_2, K) = 0$ at a point $E(K) = (\overline{x}(K), \overline{y}_1(K), \overline{y}_2(K))$ in the positive octant (see Fig. 2.4.2).

The generalization of the Rosenzweig-MacArthur graphical criterion is expressed by the following theorem (Farkas-Freedman 1989a; 1989b).

THEOREM 2.4.1. *If the equilibrium $E(K)$ of system (2.4.8) is in the interior of the positive octant and $F'_x \le 0$ at $E(K)$ then $E(K)$ is asymptotically stable.*

On Fig. 2.4.2 a typical onionlike prey zero isocline surface is shown (it can be compared to the prey isoclines on Fig. 2.2.3). Theorem 2.4.1 can still be considered a graphical criterion. Function F is increasing as we cross the surface inwards and therefore its gradient points inward. Hence, the theorem says that if the equilibrium point is on the "northern hemisphere of the onion" then it is asymptotically stable.

PROOF. If we linearize the system at equilibrium $E(K)$ then the characteristic equation turns out to be

$$\begin{aligned} &\lambda^3 + \lambda^2\left(-xF'_x - y_1G'_{1y_1} - G'_{2y_2}\right) \\ &\quad +\lambda\left(xF'_x\left(y_1G'_{1y_1} + y_2G'_{2y_2}\right) + y_1y_2G'_{1y_1}G'_{2y_2} - xy_1F'_{y_1}G'_{1x} - xy_2F'_{y_2}G'_{2x}\right) \\ &\quad +xy_1y_2\left(F'_{y_1}G'_{1x}G'_{2y_2} + F'_{y_2}G'_{2x}G'_{1y_1} - F'_xG'_{1y_1}G'_{2y_2}\right) = 0 \end{aligned} \tag{2.4.12}$$

where all functions are to be evaluated at $E(K) = (\overline{x}(K), \overline{y}_1(K), \overline{y}_2(K))$. Conditions (2.4.9)-(2.4.11) and the condition of the theorem imply that all the coefficients of the characteristic polynomial on the left-hand side of Eq. (2.4.12) are positive. The Routh-Hurwitz criterion of stability (A1.1.1) requires that

$$\begin{aligned} \left(-y_1G'_{1y_1} - y_2G'_{2y_2}\right)\left(x^2F'^2_x + y_1y_2G'_{1y_1}G'_{2y_2}\right) + xy_1^2F'_{y_1}G'_{1x}G'_{1y_1} \\ +xy_2^2F'_{y_2}G'_{2x}G'_{2y_2} \end{aligned}$$

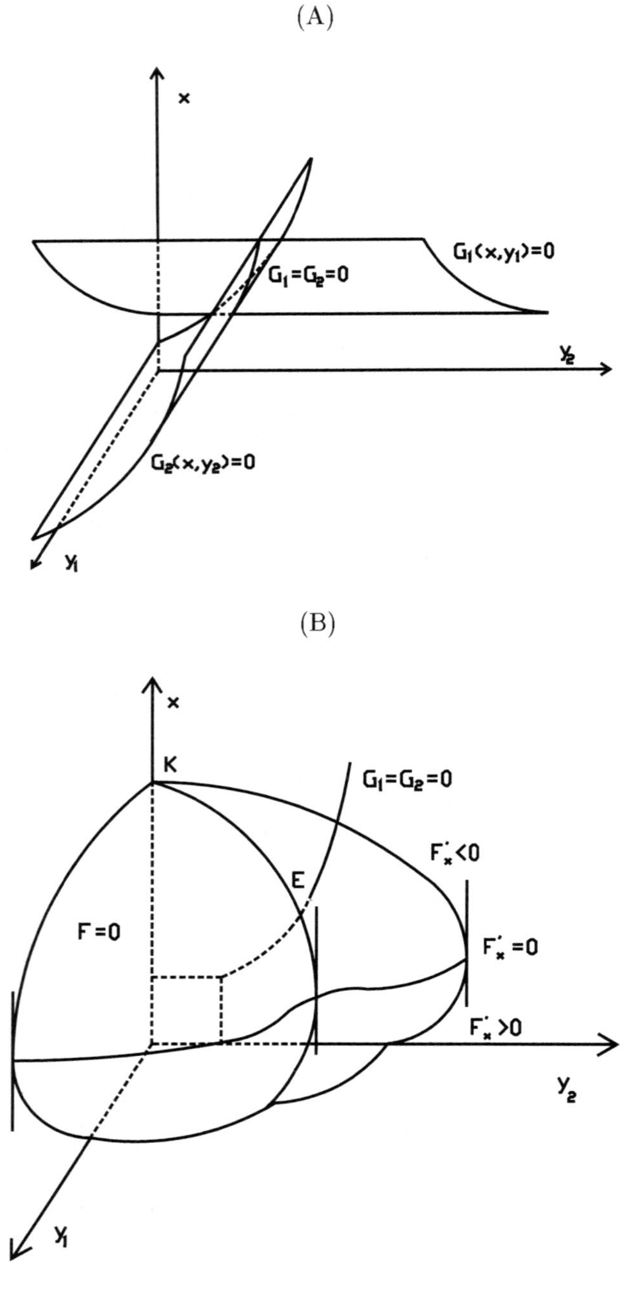

Figure 2.4.2: The equilibrium E of system (2.4.8) in a stable position. (A) The predator isoclines; and (B) the prey isocline.

$$> xF'_x\left(\left(y_1G'_{1y_1}+y_2G'_{2y_2}\right)^2-\left(xy_1F'y_1G'_{1x}+xy_2F'_{y_2}G'_{2x}\right)\right) .$$

Now, the conditions imply that the left-hand side is positive and if $F'_x \le 0$ then the right-hand side is either negative or zero, which proves the theorem. □

Theorem 2.4.1 remains true even if more than two predators are competing for a single prey species. We show this not for the sake of *l'art pour l'art* mathematical generalizations but in order to give an example for the application of the important concept of *sign stability* (see Appendix 1.1). We consider a system similar to (2.4.8) with the difference that now $n \ge 2$ predators are competing for prey:

$$\dot{x} = xF\left(x, y_1, y_2, \ldots, y_n, K\right), \quad \dot{y}_1 = y_1G_1\left(x, y_1\right), \ldots, \quad \dot{y}_n = y_nG_n\left(x, y_n\right) \tag{2.4.13}$$

where we assume that

$$(x-K)F(x,0,\ldots,0,K) < 0, \quad x \ge 0, \quad x \ne K > 0\,, \tag{2.4.14}$$

(in the absence of predation prey grows until it reaches carrying capacity),

$$F'_{y_i}\left(x, y_1, \ldots, y_n, K\right) < 0, \qquad (i = 1, \ldots, n)\,, \tag{2.4.15}$$

(the y-s are predators of x),

$G_i(0, y_i) < 0$, $G'_{iy_i}(x, y_i) \le 0$ but at most one of them is zero,

$$G'_{ix}\left(x, y_i\right) > 0\,, \qquad (i = 1, 2, \ldots, n)\,, \tag{2.4.16}$$

(predators cannot survive without prey; there is intraspecific competition within at least $n-1$ predator species; and the increase of prey is increasing the growth rate of the predators). While in the case of system (2.4.8) one may see intuitively that for reasonable functions F and G the system has an equilibrium in the positive orthant, this is now not as obvious. In Farkas (1988) it is proved that for a fairly wide class of systems (2.4.13) this is true. Assume that system (2.4.13) has an equilibrium $E(K) = (\overline{x}(K), \overline{y}_1(K), \ldots, \overline{y}_n(K))$ in the positive orthant of the $x, y_1, \ldots, y_n$ space.

THEOREM 2.4.2. *If the equilibrium $E(K)$ is in the interior of the positive orthant and $F'_x \le 0$ at $E(K)$ then $E(K)$ is asymptotically stable.*

PROOF. By linearizing the system at the equilibrium point we obtain the following $(n+1)$ by $(n+1)$ coefficient matrix where all the entries are to be taken at $E(K)$,

$$\begin{bmatrix} xF'_x & xF'_{y_1} & xF'_{y_2} & \cdots & xF'_{y_n} \\ y_1G'_{1x} & y_1G'_{1y_1} & 0 & \cdots & 0 \\ y_2G'_{2x} & 0 & y_2G'_{2y_2} & \cdots & 0 \\ \cdot & \cdot & \cdot & \cdots & 0 \\ y_nG'_{nx} & 0 & 0 & \cdots & y_nG'_{ny_n} \end{bmatrix}$$

Our assumptions imply that the first entry in the first row (column) is nonpositive, the rest of the entries in the first row are negative, the rest of the entries in the first column are positive, and the rest of the entries in the main diagonal are negative except, perhaps, one which is then zero. As a consequence, there is no cycle of length longer than two in the directed graph attached to this matrix, and as the undirected graph consists of the edges $[1, 2], [1, 3], [1, 4], \ldots, [1, n+1]$ the rest of the conditions of Theorem A1.1.4 are also fulfilled. Thus, this matrix is sign stable and this proves Theorem 2.4.2. □

One may consider as problems *dual* to (2.4.8), and (2.4.13), respectively, the case in which a single predator is consuming, respectively, two, or more than two different prey species. See Farkas (1988; 1990). Only the case where there is no competition between prey species will be considered here. The quantities of the n prey species and the predator at time t will be denoted by $x_i(t)$, $(i = 1, 2, \ldots, n)$, $y(t)$, respectively. The system is now

$$\dot{x}_i = x_i F_i (x_i, y, K_i), \quad (i = 1, 2, \ldots, n), \quad \dot{y} = yG(x_1, x_2, \ldots, x_n, y), \tag{2.4.17}$$

where

$$(x_i - K_i) F_i (x_i, 0, K_i) < 0, \qquad x_i \geq 0, \quad x_i \neq K_i > 0, \quad (i = 1, 2, \ldots, n) \tag{2.4.18}$$

(each prey species can grow up to its own carrying capacity),

$$F'_{iy} < 0, \quad G(0, \ldots, 0, y) < 0, \quad G'_{x_i} > 0, \quad G'_y \leq 0, \quad (i = 1, 2, \ldots, n) \tag{2.4.19}$$

(y is predator for each prey; the predator cannot survive without prey; the increase of each prey species is advantageous for the predator; and there may be intraspecific competition within the predator species).

One may prove again that a fairly large class of systems (2.4.17) has an equilibrium point $E(K) = (\overline{x}_1(K), \overline{x}_2(K), \ldots, \overline{x}_n(K), \overline{y}(K))$ in the positive orthant of the $x_1, x_2, \ldots, x_n, y$ space (K is shorthand here for $(K_1, K_2, \ldots, K_n)$). Linearizing the system at the equilibrium and establishing, as in the preceding, the sign stability of the coefficient matrix of the linearized system one can prove the analog of the Rosenzweig-MacArthur graphical criterion for this case:

THEOREM 2.4.3. *If the equilibrium $E(K)$ of system (2.4.17) is in the interior of the positive orthant,*

$$F'_{ix_i}(\overline{x}_i(K), \overline{y}(K), K_i) \leq 0, \qquad (i = 1, 2, \ldots, n), \tag{2.4.20}$$

but at most one of these partial derivatives is zero, then the equilibrium is asymptotically stable.

The previous problem leads us to the ecologically meaningful and important question of whether two competing species that would outcompete each other could be made to coexist by the introduction of a predator. This problem and its generalizations have attracted much attention in the past 25 years. The problem is usually treated in the context of *permanence* of ecological systems. The theory of permanent coexistence of ecological communities has a vast literature (see, e.g., Hofbauer and Sigmund 1988, and the references therein) and there are several definitions whose hierarchy is discussed in Freedman and Moson (1990). The point is that one is not often interested in whether the species live together in a state of equilibrium or in a cyclic, periodic way or even such a way that their quantities vary chaotically; instead, the major question is whether *all the species will survive in the long run.* This last important property is ensured if the coexistence is global, which means that no matter what positive initial values the quantities of the species in the system have they will continue to coexist, and if the trajectories describing the dynamics of the system stay near to the boundary of the positive orthant but for a finite time. These requirements are important because sudden changes in the numbers may move the system out from the basin of an asymptotically stable equilibrium, say, and throw it into the basin of an attractor in the boundary. The result would be the extinction of some of the species and even small stochastic fluctuations of the numbers make it possible that species will be wiped out if the trajectory stays near to the boundary for a long time. From this point of view neither a *locally* asymptotically stable equilibrium nor an attractor like L of system (2.4.6) is satisfactory. We accept the following definition (see Schuster, et al., 1979; and Hutson and Moran, 1982).

DEFINITON 2.4.1. We say that system (2.4.1) is *permanent* if there is a compact (closed and bounded) set K in the interior of the positive orthant of the phase space such that to every initial value with all the coordinates positive there belongs a time $T > 0$ such that the corresponding trajectory enters K and stays in it for $t > T$.

Because the distance of the compact set K from the boundary is positive, say $\delta > 0$, permanence means that no matter which positive initial values are given the corresponding trajectory will not stay closer to the boundary of the positive orthant than δ for more time than T (depending on the initial values). In other words, the boundary is a *repeller* of the system. Note that permanence also means that every solution belonging to initial conditions in the interior of the positive orthant is bounded.

For conditions of permanence see Hofbauer (1981).

We turn now to the problem of stabilization of a competitive system by the introduction of predators. Following the fundamental paper by Hutson and Vickers (1983), we consider here a Lotka-Volterra system of two competitive preys and a single predator :

$$\begin{aligned} \dot{x}_1 &= x_1(r_1 - a_{11}x_1 - a_{12}x_2 - a_{13}y) \\ \dot{x}_2 &= x_2(r_2 - a_{21}x_1 - a_{22}x_2 - a_{23}y) \\ \dot{y} &= y(-c + \beta_1 x_1 + \beta_1 x_2 - a_{33}y)\ . \end{aligned} \tag{2.4.21}$$

It is assumed that

$$a_{12}, a_{21}, a_{33} \geq 0 \quad \text{and the rest of the coefficients are positive,} \tag{2.4.22}$$

which means that there might be interspecific competition between the prey species, the per capita mortality of the predator may increase with the numbers and otherwise, x_1 and x_2 are the quantities of the prey species and y is that of the predator. (As compared to (2.4.2) $r_3 = -c, a_{31} = -\beta_1, a_{32} = -\beta_2$.) In case $y \equiv 0$ we get back the 2D competitive system (2.3.1) studied in the previous Section. Hutson and Vickers prove that if conditions (2.4.22) hold then in order to have permanence it is necessary that the system have a unique equilibrium in the interior of the positive octant and they prove the following theorem.

THEOREM 2.4.4. *Suppose that conditions (2.4.22) hold and the system has a unique equilibrium in the interior of the positive octant; if the determinant D of the community matrix*

$$A = \begin{bmatrix} a_{11} & a_{12} & a_{13} \\ a_{21} & a_{22} & a_{23} \\ -\beta_1 & -\beta_2 & a_{33} \end{bmatrix}$$

is positive and either $r_1 a_{22} > r_2 a_{12}$ or $r_2 a_{11} > r_1 a_{21}$ then system (2.4.21) is permanent; if either $D < 0$ or both previous inequalities are reversed, $r_1 a_{22} < r_2 a_{12}$ and $r_2 a_{11} < r_1 a_{21}$, then the system has an asymptotically stable equilibrium on the boundary of the positive octant and it is not permanent.

If there is a unique equilibrium in the interior of the positive octant then D cannot be zero, so that apart from nongeneric equalities this theorem settles the problem of permanence for the system. The determinant of the community matrix A is

$$D = a_{33}(a_{11}a_{22} - a_{12}a_{21}) + a_{23}(\beta_2 a_{11} - \beta_1 a_{12}) + a_{13}(\beta_1 a_{22} - \beta_2 a_{21}) .$$

Comparing the conditions of the theorem with cases (A), (B), and (C) of system (2.3.1) it is fairly easy to see that in cases (A) and (C) one may find a predator with suitable parameters that makes the system permanent while in case (B) no such predator can be found. This means that if one of the prey species outcompeted the other one independently of the initial values (case (C)) then a suitable predator may ensure the survival of both species but a bistable system (case (B)) cannot be made permanent by introducing a predator.

There are several generalizations of the previous result. Kirlinger (1986) has proved, for example, that under certain conditions even a bistable competitive two-species Lotka-Volterra system ((2.3.1) case (B)) can be made permanent by the introduction of two predators each feeding on just one (separate) prey.

2.5 More Realism

In this Section tools will be considered by which more realistic models can be produced that take into account delay, the age structure of the populations, and spatial dispersion. In order to take into account these effects one has to go beyond the ordinary and apply delay and partial differential equations. Because both delay and partial differential equations are more difficult to handle than ordinary ones, the question arises as is always the case with mathematical modeling, how far is it worthwhile to go in approximating reality? The more complicated the models we use the fewer conclusions we may deduce from them while from simple models we may draw more conclusions. However, do our conclusions then reflect reality? In this Section some examples will be given showing how delay, age structure, and spatial distribution of populations can be built into our models.

Delay. Delay plays an important role in the dynamics of populations. The offspring appears but must reach maturity before taking part in the reproduction process and the living creature must eat in order to be able to reproduce but the offspring appear only after some time (the gestational period, say, in the case of mammals). We present here a predator prey model in which the growth rate of the predator does not depend on only the present quantity of prey but on the prey that was available in the past. We shall show how the delayed effect of the past, or rather the increase of this delay, influences the behavior of the system.

We start with system (2.2.1) where $N(t)$, $P(t)$ denote the quantities of prey and predator, respectively, at time t, the parameters $\varepsilon, \alpha, \gamma, \beta, K > 0$, $\delta = 0$, that is, predator mortality is constant but we take into account that the predator growth rate at present depends on past quantities of prey and therefore a continuous *weight* (or *density*) *function* G is introduced whose role is to weigh moments of the past. Function G satisfies the requirements

$$G(s) \geq 0\,, \quad s \in (0, \infty)\,, \quad \int_0^\infty G(s)\,\mathrm{d}s = 1\,, \tag{2.5.1}$$

and $N(t)$ is replaced in the second equation by its weighted average over the past:

$$Q(t) := \int_{-\infty}^{t} N(\tau) G(t-\tau)\,\mathrm{d}\tau\,. \tag{2.5.2}$$

This way we arrive at the delay system

$$\begin{aligned} \dot{N}(t) &= \varepsilon N(t)(1 - N(t)/K - P(t)\alpha/\varepsilon)\,, \\ \dot{P}(t) &= -\gamma P(t) + \beta P(t) \int_{-\infty}^{t} N(\tau)\, G(t-\tau)\,\mathrm{d}\tau\,. \end{aligned} \tag{2.5.3}$$

Here the delay is *continuous*, meaning that the quantity of prey has an influence on the present growth rate of predator not just at a single moment in the past but over the whole past, or at least in those time intervals where the weight function G is not zero. The weighted time average Q of N is between the maximal and minimal values of N and in case N was constant over the whole past, Q is equal to the same constant. Several authors have dealt with system (2.5.3), including Cushing (1977), Farkas (1984a), Szabo (1987), Farkas et al. (1988) etc. It is easy to see that the system has three equilibria: $(0,0)$, $(K,0)$ which represents the absence or the dying out of the predator, and $(\overline{N},\overline{P}) = (\gamma/\beta, (1-\gamma/(K\beta))\varepsilon/\alpha)$, with the latter in the interior of the positive quadrant of the N, P plane iff

$$0 < \gamma < K\beta\,. \tag{2.5.4}$$

(Cf. (2.2.2) and the explanation given for this condition there). The problems arise when we want to check the stability of the equilibria because a typical delay differential equation has an infinite spectrum (has an infinite number of eigenvalues, see Appendix 2.2). In order to handle this situation one has to specify the weight function G and methods worked out for functional differential equations are to be applied (see Stépán 1986). However, if G is of a special form, namely, a polynomial times exponential function, then the system is equivalent to a higher dimensional system of ordinary differential equations (see Fargue, 1973). The simplest choice is $G(s) = a\exp(-as)$, with $a > 0$. This function satisfies the conditions (2.5.1), and with this choice applying the notation (2.5.2), on $(0,\infty)$ system (2.5.3) is equivalent to the following system:

$$\begin{aligned} \dot{N} &= \varepsilon N(1-N/K) - \alpha NP \\ \dot{P} &= -\gamma P + \beta PQ \\ \dot{Q} &= a(N-Q)\,. \end{aligned} \tag{2.5.5}$$

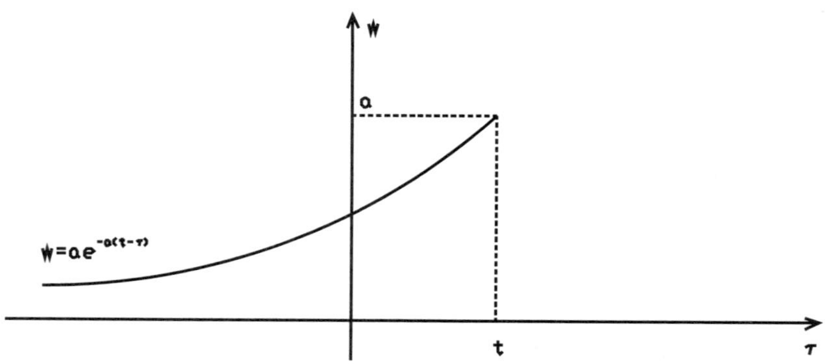

Figure 2.5.1: Exponentially fading memory in system (2.5.3).

We call this choice of G *exponentially fading memory* because the largest weight is given to moments in the neighborhood of the present (t) and as we go back in time the weight is decreasing exponentially (see Fig. 2.5.1). If we look at Fig. 2.5.1 we see that the larger a is the less important is the influence of the past, "memory fades away faster," and this *delay increases with the decrease of parameter* a. The equilibria of the latter system corresponding to those of (2.5.3) are $(0,0,0)$, $(K,0,K)$ and $E = (\overline{N},\overline{P},\overline{Q}) = (\gamma/\beta,\ (1-\gamma/(K\beta))\varepsilon/\alpha,\ \gamma/\beta)$, respectively. Linearizing system (2.5.5) at $(0, 0, 0)$ and $(K, 0, K)$ we find that the first one is always unstable and the second one is unstable if condition (2.5.4) holds and is asymptotically stable if the inequality there is reversed. As this has already been mentioned in Section 2.2, this is fairly intuitive–a predator dies out if its mortality is high, food is scarce, and the predator is converting food inefficiently. In the sequel we assume (2.5.4) so that the only possibly stable equilibrium is now E which represents the coexistence of predator and prey. The stability problem of E is a difficult one; we give a sketch here as to how it can be settled (for details, see Farkas 1994). We linearize system (2.5.5) at E and apply the Routh-Hurwitz criterion (Theorem A1.1.2 and (A1.1.1)). Considering the delay measured by $1/a$ as a "bifurcation parameter" and fixing the rest of the parameters, if $a > K\beta - \gamma - \gamma\varepsilon/\,(K\beta)$, that is, the delay is small, then E is asymptotically stable and if the inequality is reversed here then it is unstable. Now, if $a_0 := K\beta - \gamma - \gamma\varepsilon/\,(K\beta)$ is negative or zero then any delay goes and the equilibrium E stays stable; however, if

$$a_0 := K\beta - \gamma - \gamma\varepsilon/\,(K\beta) > 0\,, \tag{2.5.6}$$

that is, $K\beta$ is not only bigger than γ but even bigger than $\gamma + \gamma\varepsilon/(K\beta)$, then if the delay is increased beyond $1/a_0$ equilibrium E is destabilized. This is an example of the rule of thumb (which is far from being universally true) that delay, or rather, an increase of delay has a destabilizing effect. We assume from now on that (2.5.6) holds and raise the question as to what happens to the system if the delay is large ? By a fairly tough calculation one may show that as a is decreased through a_0 the equilibrium undergoes an Andronov-Hopf bifurcation (see Appendix 2.3) and that if

$$K\beta - \gamma - \gamma\varepsilon/(K\beta) - \varepsilon^2\gamma/(K\beta(2\gamma + \varepsilon)) > 0 \tag{2.5.7}$$

then this bifurcation is supercritical. Note that condition (2.5.7) implies (2.5.6), which in turn implies (2.5.4). This means that if condition (2.5.7) holds and the delay is larger than $1/a_0$ then predator and prey will no longer continue to coexist in an equilibrium state but will vary in numbers in a small amplitude, orbitally asymptotically stable periodic oscillation.

System (2.5.3) has been studied by Farkas and Farkas (1988) by applying the weight function $G(s) = a^2 s \exp(-as)$, $a > 0$, called a memory with a hump. This weight function has the advantage that it specifies a moment in the past when the quantity of food is the most important from the point of view of the present growth of the predator but the analysis is more difficult and requires the use of computer algebra.

Age structure. In Section 1.2 we introduced a model that takes into consideration the age distribution of the population. There this was done in the discrete setting–time and age were considered to pass in discrete units and the population was divided up into a finite number of age groups. Now we are to show how age structure can be taken into acount in the continuous setting, which is more realistic in the case of mammals for example. (For some insects it seems to be satisfactory to divide the population into two age groups, larvae and imagines.) We consider a single population and denote the density of members of age a at time t by $n(t, a)$. This means that the quantity of members between age a and age $a + \mathrm{d}a$ is $n(t, a)\mathrm{d}a$ for small $\mathrm{d}a$ and the quantity of the total population at time t is

$$N(t) = \int_0^\infty n(t, a)\,\mathrm{d}a = \int_0^{\overline{a}} n(t, a)\,\mathrm{d}a\,,$$

where $\overline{a} > 0$ is the maximal span of life for members of the given population if such a span can be specified. Now, suppose that there is no migration into and out of the population and see how the density varies over time. The quantity of members between age a and age $a + \mathrm{d}a$ at time $t + \mathrm{d}t$ is $n\,(t + \mathrm{d}t, a)\,\mathrm{d}a$; if we suppose that $\mathrm{d}t = \mathrm{d}a$ then this number is equal to the number of those who were between ages $a - \mathrm{d}a$ and a at time t minus those who died from the latter age group during the time $\mathrm{d}t$. Denoting the *mortality* of the members of age a by $\mu(a)$, the last quantity is $n\,(t, a - \mathrm{d}a)\,\mathrm{d}a\mu\,(a)\,\mathrm{d}t$, so that the balance equation is

$$n\,(t + \mathrm{d}t, a)\,\mathrm{d}a = n\,(t, a - \mathrm{d}a)\,\mathrm{d}a - n\,(t, a - \mathrm{d}a)\,\mathrm{d}a\;\mu\,(a)\,\mathrm{d}t$$

or

$$\begin{aligned} n\,(t + \mathrm{d}t, a)\,\mathrm{d}a - n\,(t, a - \mathrm{d}a)\,\mathrm{d}a &= n\,(t + \mathrm{d}t, a)\,\mathrm{d}a - n\,(t, a)\,\mathrm{d}a \\ &\quad + n\,(t, a)\,\mathrm{d}a - n\,(t, a - \mathrm{d}a)\,\mathrm{d}a \\ &= -n\,(t, a - \mathrm{d}a)\,\mathrm{d}a\;\mu\,(a)\,\mathrm{d}t\,. \end{aligned}$$

Dividing by $\mathrm{d}a\;\mathrm{d}t$ and taking into account that $\mathrm{d}a = \mathrm{d}t$, we obtain

$$\frac{n\,(t + \mathrm{d}t, a) - n\,(t, a)}{\mathrm{d}t} + \frac{n\,(t, a) - n\,(t, a - \mathrm{d}a)}{\mathrm{d}a} = -\mu\,(a)\,n\,(t, a - \mathrm{d}a)\;,$$

or taking the limits as $\mathrm{d}a = \mathrm{d}t$ tends to zero

$$\frac{\partial n}{\partial t} + \frac{\partial n}{\partial a} = -\mu\,(a)\,n\,. \tag{2.5.8}$$

As migration has been excluded, one may enter the population only by being born in it at the age of $a = 0$. We denote the *fertility* of those of age a by $b(a)$; this means that the contribution of the age group between a and $a + \mathrm{d}a$ to the newly born offspring at time t is $b(a)n(t, a)\mathrm{d}a$ for small $\mathrm{d}a$. As a consequence

$$n\,(t, 0) = \int_0^\infty b\,(a)\,n\,(t, a)\;\mathrm{d}a = \int_0^{\overline{a}} b\,(a)\,n\,(t, a)\;\mathrm{d}a\,. \tag{2.5.9}$$

The first order linear partial differential equation (2.5.8) and the integral equation (2.5.9) together play the role of Eq. (1.2.1) of the discrete case. While for that case one had to prescribe an initial population vector, here we need an initial age distribution

$$\nu\,(a) = n\,(0, a)\;. \tag{2.5.10}$$

Equations (2.5.8)–(2.5.10) determine the dynamics of the age-structured population for $t > 0$. Clearly, mortality μ, fertility b, and the initial age distribution ν are nonnegative functions defined on $(0, \infty)$ or on $(0, \overline{a})$ and ν has to satisfy the compability condition

$$\nu(0) = n(0,0) = \int_0^{\infty} b(a)\nu(a)da = \int_0^{\overline{a}} b(a)\nu(a)da \,.$$

It is hopeless to attempt an analytical solution of the problem for arbitrarily (but reasonably) given functions μ, b, and ν. There are two things we can do and we shall treat this problem in somewhat more detail than for other problems discussed in this book because it is difficult to give a reference that explains the study.

First we try to solve Eq. (2.5.8) by the method of characteristics (see Appendix 3.1). The ODE system of characteristics is

$$dt/d\tau = 1 \,, \qquad da/d\tau = 1 \,, \qquad dn/d\tau = -\mu\,(a)\,n \,. \tag{2.5.11}$$

As we know that for $t = 0$, $a = a_0$ we have $n(0, a_0) = \nu\,(a_0)$, we choose as initial conditions $(t\,(0)\,, a\,(0)\,, n\,(0)) = (0, a_0, \nu\,(a_0))$. The corresponding solution is

$$t = \tau, \quad a = \tau + a_0 = t + a_0, \quad n = \nu(a_0)\exp\left(-\int_0^t \mu(s + a_0)\,\mathrm{d}s\right) \,.$$

Substituting from the second relation $a_0 = a - t$ into the third one we obtain as a solution

$$n(t,a) = \nu(a-t)\exp\left(-\int_0^t \mu(s + a - t)\,\mathrm{d}s\right) \,. \tag{2.5.12}$$

Although this is a solution, the trouble with it is that because μ and ν are defined only for nonnegative arguments this solution is defined only for $a \geq t$, that is, for ages that are higher than the actual time from the start, or in other words, this solution yields only the time evolution of that part of the population that was present at the start. This is clear also because the solution does not depend on the birth process b, that is, it does not incorporate Eq. (2.5.9). In order to solve system (2.5.11) in the domain $t \geq a$ we use Eq. (2.5.9) as initial conditions, that is, we choose as initial values $(t\,(0)\,, a\,(0)\,, n\,(0)) = (t_0, 0, n\,(t_0, 0))$ where

$$n\,(t_0, 0) = \int_0^{\infty} b\,(a)\,n\,(t_0, a)\,\mathrm{d}a = \int_0^{\overline{a}} b\,(a)\,n\,(t_0, a)\,\mathrm{d}a \,. \tag{2.5.13}$$

We obtain for the characteristics

$$t = \tau + t_0 \,, \quad a = \tau \,, \quad n = n(t_0, 0)\exp\left(-\int_0^{\tau} \mu(s)\,\mathrm{d}s\right) \,. \tag{2.5.14}$$

By substituting from the first and second equation $t_0 = t - a$ into the third one and applying Eq. (2.5.13) we obtain the solution

$$\begin{aligned} n(t,a) &= n(t-a,0)\exp\left(-\int_0^a \mu(s)\,\mathrm{d}s\right) \\ &= \int_0^\infty b(s)\,n(t-a,s)\,\mathrm{d}s\ \exp\left(-\int_0^a \mu(s)\,\mathrm{d}s\right) \\ &= \int_0^{\overline{a}} b(s)\,n(t-a,s)\,\mathrm{d}s\ \exp\left(-\int_0^a \mu(s)\,\mathrm{d}s\right) \end{aligned} \tag{2.5.15}$$

for $t \geq a$. The last expression is naturally not an explicit solution but an integral equation to be solved. However, the solution obtained for $a \geq t$ and the last one fit together when $a = t$:

$$\nu(0)\exp\left(-\int_0^t \mu(s)\,\mathrm{d}s\right) = n(t,t) = \int_0^\infty b(s)n(0,s)\,\mathrm{d}s\ \exp\left(-\int_0^a \mu(s)\,\mathrm{d}s\right) .$$

Equation (2.5.15) shows that in order to determine the quantity of those of age a at time t we must know the age distribution at time $t - a$. Naturally, $t - a$ may be smaller than the maximal age $\overline{a}$ and, as a consequence, in the integral s assumes values larger than $t - a$; in the latter domain we must substitute the solution from Eq. (2.5.12). We do not pursue methods to handle (mainly in an approximative way) the determination of the solution in the general case (see e.g., Metz and Diekmann, 1986) but turn now to the special problem of solutions that reflect an age distribution that is constant over time (the dream of every demographer and minister of education).

We look for a solution of the form

$$n(t,a) = \nu(a)\exp(\gamma t) ,$$

that is, the total quantity of the population may increase ($\gamma > 0$), decrease ($\gamma < 0$), or stay constant ($\gamma = 0$) in time but the relative age distribution given by the function ν remains the same. By substituting this solution into Eq. (2.5.8) we obtain for the age distribution $\nu(a)$ an ordinary differential equation

$$\nu'(a) = -(\gamma + \mu(a))\nu(a) \tag{2.5.16}$$

that is easy to solve. The solution is

$$\nu(a) = \nu(0)\exp\left(-\int_0^a (\gamma + \mu(s))\,\mathrm{d}s\right) . \tag{2.5.17}$$

We see that constant γ and mortality function μ determine the initial stationary age distribution apart from the number of newborns $\nu(0)$ at time $t = 0$. Substituting this solution into Eq. (2.5.9) we get the condition

$$1 = \int_0^{\overline{a}} b(a)\exp(-\gamma a)\exp\left(-\int_0^a \mu(s)\,\mathrm{d}s\right)\mathrm{d}a \tag{2.5.18}$$

for the birth rate b. From now on we work with the maximal life span $\overline{a}$ instead of infinity. The last condition does not determine b uniquely. It is more realistic to assume that mortality and birth rate are given. The last condition determines then the constant γ uniquely, because one can easily show that the right-hand side of Eq. (2.5.18) is a decreasing function of γ, it assumes large positive values for in modulus large negative γ, and tends to zero as γ tends to infinity. This way the stationary age distribution is determined uniquely by Eq. (2.5.17). If $\gamma = 0$ is substituted into the right-hand side of Eq. (2.5.18) then we get

$$K := \int_0^{\overline{a}} b(a) \exp\left(- \int_0^a \mu(s)\, \mathrm{d}s \right) \mathrm{d}a\,.$$

The constant γ that satisfies Eq. (2.5.18) is, respectively, positive and negative according to whether K is greater or less than 1. Thus, the value of K determines whether the population of constant relative age distribution is growing or decreasing in numbers.

A numerical example for human population will be given. The maximal life span $\bar{a}$ will be taken as 100 years or 10 decades. According to Tusnády (1995), if age is measured in *decades* (10 years) the probability that a person of age a will die within 1 year is the reciprocal of the exponential of the difference between the maximal life span and his/her age (both measured in decades). The law applies between the ages 20 (or 2 decades) and 90 (or 9 decades). However, disregarding infant mortality we set

$$\mu(a) = \frac{1}{\exp(10-a)}\,, \qquad 0 < a < 10\,,$$

(see Fig. 2.5.2 (A)). We assume birth rate in such a way that it is zero below the age of 1 decade and above the age of 9 decades and has a maximum at an age higher than 2 decades and less than 3 decades. A suitable choice is (see Fig. 2.5.2 (B))

$$b(a) = \begin{cases} 0, & 0 < a < 1 \\ 0.061(a-1)(9-a)\exp(-(a-2.5)^2), & 1 < a < 9 \\ 0, & 9 < a < 10 \end{cases}\,.$$

Integration by MAPLE-V yields $K = 1$, hence $\gamma = 0$. Substituting into Eq. (2.5.17) we obtain for the stationary age distribution (see Fig. 2.5.2 (C))

$$\nu(a) = \nu(0) \exp(0.000045(1 - e^a))\,.$$

As $\gamma = 0$, this is at the same time the stationary (time independent) solution.

Spatial distribution. In the models considered up to this point it was tacitly assumed that members of the same or different populations may meet any other member with equal probability. This assumption plays an important role in problems concerning reproduction and also in predator-prey and other relations. This means that populations or ecological systems are considered

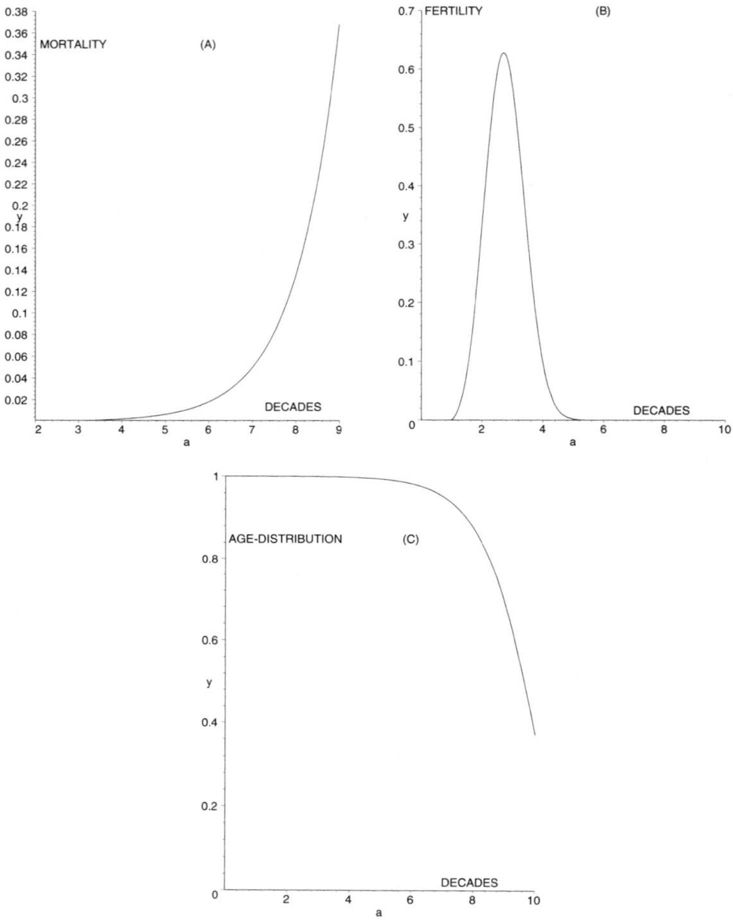

Figure 2.5.2: Mortality (A), fertility (B), and age distribution (C) for the example of constant age distribution; the horizontal axis is age measured in decades (MAPLE-V).

either to be concentrated in a point of space or they are "well stirred." This assumption is acceptable if the system is that of a small lake, a small wood, or a test tube in a laboratory. However, one generally has to take into consideration that populations must exploit a smaller or larger territory on the surface of the Earth or the sea, and the distance between the members may count in possible interactions. Thus, we consider populations that live in a 3-, 2-, or 1D space, for instance fish in the sea or in a lake, animals on the savannah, or worms moving up and down on a single tree. Sometimes we substitute the actual physical 3-, or 2D space by a space, which clearly renders the results more abstract but simplifies the mathematics. Thus, rather than population quantities we deal instead with the spatial densities (quantities per unit volume, area, or length) of the populations. It is usually assumed that populations diffuse away from places of high density towards places of small density following Fick's law, that is, in the direction of the negative gradient of the density and with a speed that is proportional to the modulus of the gradient (see Appendix 3.2). This way the dynamics is described by a system of reaction-diffusion equations, that is, a nonlinear or rather quasilinear system of parabolic partial differential equations (Appendix 3.2). In order to illustrate some of the problems, phenomena, and methods that arise we present here a 2D predator-prey system (see Cavani and Farkas, 1994). More can be found in Murray (1989), Okubo (1980) and Czárán (1998).

Consider a prey species that in the absence of predation follows the logistic dynamics and a predator that consumes the prey by a Holling type functional response (cf. system (2.4.3)) and whose death rate is not constant but instead increases when there is overcrowding. The two species live together in the same territory, which is sealed off so that there is no migration in or out (for example, an island) but they difffuse freely within. The territory will be represented by a bounded interval $0 \leq x \leq l$. The densities of prey and predator as functions of time t and the spatial coordinate x will be denoted by $N(t,x)$ and $P(t,x)$, respectively. The system that describes the dynamics is then

$$\begin{aligned} \partial N/\partial t &= d_N \partial^2 N/\partial x^2 + \varepsilon N(1 - N/K) - \beta NP/(\beta + N) \\ \partial P/\partial t &= d_P \partial^2 P/\partial x^2 - P(\gamma + \delta P)/(1+P) + \beta NP/(\beta + N) \end{aligned} \qquad (2.5.19)$$

where $0 \leq t < \infty, \quad 0 \leq x \leq l$ and all the constants $\varepsilon, \beta, \gamma, \delta, K$ and the diffusion coefficients d_N, d_P are positive. *No-flux boundary conditions* are attached to the problem

$$N_x'(t,0) = N_x'(t,l) = P_x'(t,0) = P_x'(t,l) = 0\,. \qquad (2.5.20)$$

If there is no diffusion (the diffusion coefficients are zero) the system of partial differential equations (PDE) (2.5.19) reduces to a system of ordinary differential equations (ODE),

$$\begin{aligned} \dot{N} &= \varepsilon N(1 - N/K) - \beta NP/(\beta + N) \\ \dot{P} &= -P(\gamma + \delta P)/(1+P) + \beta NP/(\beta + N)\,, \end{aligned} \qquad (2.5.21)$$

which is the *kinetic system* attached to (2.5.19). We study first this kinetic system. The equation of the prey zero isocline is

$$P = H_1(N) := (K - N)(\beta + N)\varepsilon/(\beta K)$$

and that of the predator is

$$P = H_2(N) := ((\beta - \gamma)N - \beta\gamma)/((\delta - \beta)N + \beta\delta)\,.$$

The equilibria of (2.5.21) are $(0,0)$, $(K,0)$ and if there is any other equilibrium then its coordinates $(\overline{N}, \overline{P})$ satisfy $\overline{P} = H_1(\overline{N}) = H_2(\overline{N})$. Now we assume

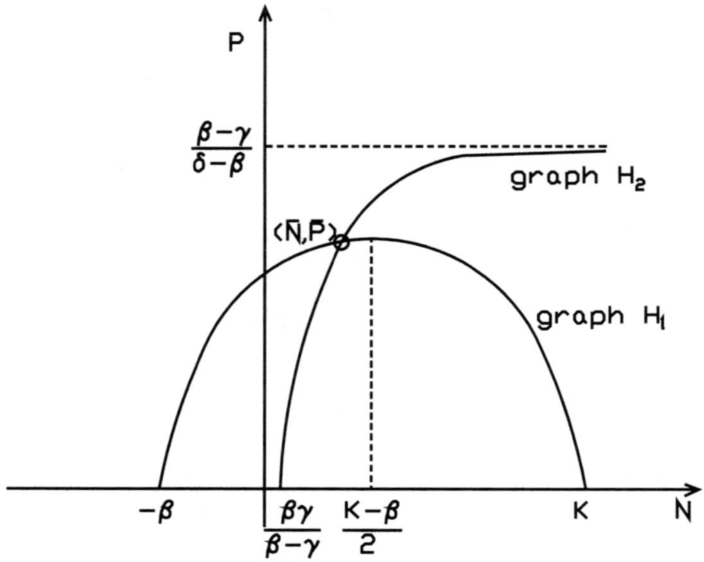

Figure 2.5.3: Prey and predator zero isoclines of system (2.5.21); $(\overline{N}, \overline{P})$ in the Allée-effect zone.

that

$$\gamma < \beta \leq \delta \quad , \quad \beta < K \quad , \quad \gamma < \beta K / (\beta + K) \ . \tag{2.5.22}$$

The first condition ensures that predator mortality increases with quantity and that the predator zero isocline has a reasonable concave down shape. The second condition ensures that the prey has an Allée-effect zone (cf. the discussion of system (2.2.4)) where the increase in prey density is favorable for its growth rate. The third condition ensures the existence of at least one positive equilibrium, denoted by $(\overline{N}, \overline{P})$, see (Fig. 2.5.3). Linearizing system (2.5.21) at the equilibria it turns out that under conditions (2.5.22) the origin and $(K, 0)$ are unstable while $(\overline{N}, \overline{P})$ is asymptotically stable if it is outside the Allée-effect zone, that is, right to the maximum point of the prey isocline. If $(\overline{N}, \overline{P})$ is at the maximum point it is still linearly asymptotically stable and, as the conditions depend continuously on the position of the equilibrium, it remains asymptotically stable even if it is left of the maximum but sufficiently near to it. Any constant solution of (2.5.21) is obviously also a solution of the PDE system (2.5.19), satisfying also the boundary conditions (2.5.20) and if such an equilibrium is unstable for (2.5.21) then it is also unstable with respect to the problem (2.5.19)–(2.5.20). The question naturally arises as to whether an asymptotically stable solution of (2.5.21) is also necessarily an asymptotically stable solution of the problem (2.5.19)–(2.5.20), meaning that all solutions of the problem with initial values near to the equilibrium tend to it as t tends to infinity? One would expect a positive answer to this question because diffusion *ususally* has a stabilizing effect; it smoothes out differences. We shall see that this is not always the case; in this model there are domains in the parameter space where although $(\overline{N}, \overline{P})$ is asymptotically stable with respect to system (2.5.21), it is unstable with respect to the problem (2.5.19)–(2.5.20); this phenomenon is called *Turing* (or *diffusion driven*) *instability* (see Appendix 3.3).

The linearization of system (2.5.21) at $(\overline{N}, \overline{P})$ yields the coefficient matrix

$$A = \begin{bmatrix} \eta\Theta_1\Theta_2 & -\Theta_1 \\ \beta^2\eta\Theta_3 & -\eta\Theta_3\Theta_4 \end{bmatrix}$$

where

$$\begin{aligned} \eta &= \varepsilon/(K\beta), \quad \Theta_1 = \beta\overline{N}/(\beta + \overline{N}), \quad \Theta_2 = K - \beta - 2\overline{N} \\ \Theta_3 &= (K - \overline{N})/(\beta + \overline{N}), \quad \Theta_4 = ((\delta - \beta)\overline{N} + \beta\delta)^2/(\delta - \gamma). \end{aligned}$$

By our assumptions all of these parameters are positive except, possibly, Θ_2, which is positive if the equilibrium point is in the Allée-effect zone and negative if it is to the right of the maximum point of the prey isocline. The linearization of the PDE system (2.5.19) with notations $v_1 = N - \overline{N}$, $v_2 = P - \overline{P}$ leads to

$$\frac{\partial}{\partial t}\begin{bmatrix} v_1 \\ v_2 \end{bmatrix} = D\frac{\partial^2}{\partial x^2}\begin{bmatrix} v_1 \\ v_2 \end{bmatrix} + A\begin{bmatrix} v_1 \\ v_2 \end{bmatrix} \quad \text{where} \quad D = \begin{bmatrix} d_N & 0 \\ 0 & d_P \end{bmatrix} \tag{2.5.23}$$

with boundary conditions similar to (2.5.20). This linear approximation of the original nonlinear system is solved by the method of separation of variables. According to this the solution is assumed in the form

$$\begin{bmatrix} v_1(t,x) \\ v_2(t,x) \end{bmatrix} = \psi(x)\begin{bmatrix} y_1(t) \\ y_2(t) \end{bmatrix}.$$

The functions of a single variable satisfy

$$\frac{\mathrm{d}}{\mathrm{d}t}\begin{bmatrix} y_1 \\ y_2 \end{bmatrix} = (A - \lambda D)\begin{bmatrix} y_1 \\ y_2 \end{bmatrix} \tag{2.5.24}$$

and

$$\mathrm{d}^2\psi/\mathrm{d}x^2 = -\lambda\psi, \qquad \psi'(0) = \psi'(l) = 0, \tag{2.5.25}$$

respectively. One first solves the boundary value problem (2.5.25), which has solutions only if

$$\lambda = \lambda_j = (j\pi/l)^2, \quad j = 0, 1, 2, \ldots. \tag{2.5.26}$$

If the jth *eigenvalue* λ_j is substituted into (2.5.25), the solution, the corresponding *eigenfunction* will be

$$\psi_j(x) = \cos(j\pi x/l).$$

Having substituted this λ_j into (2.5.24), we solve that system and get two independent solutions. Multiplying these with $\psi_j(x) = \cos(j\pi x/l)$ and summing up for j we obtain the "general solution" of the boundary value problem attached to the linear system (2.5.23),

$$\begin{bmatrix} v_1(t,x) \\ v_2(t,x) \end{bmatrix} = \sum_{j=0}^{\infty} \left(a_{1j} \begin{bmatrix} y_{11j}(t) \\ y_{21j}(t) \end{bmatrix} + a_{2j} \begin{bmatrix} y_{12j}(t) \\ y_{22j}(t) \end{bmatrix} \right) \cos(j\pi x/l) \qquad (2.5.27)$$

where the coefficients a_{1j}, a_{2j} are to be determined according to the initial function prescribed at $t = 0$:

$$\begin{bmatrix} V_1(x) \\ V_2(x) \end{bmatrix} = \begin{bmatrix} v_1(0,x) \\ v_2(0,x) \end{bmatrix}.$$

Now, if for every λ_j $(j = 0, 1, 2, \ldots)$ system (2.5.24) is asymptotically stable, then each coordinate function y in the series (2.5.27) tends to zero exponentially as t tends to infinity and this implies that the constant solution $(\overline{N}, \overline{P})$ of the boundary value problems (2.5.19)–(2.5.20) is asymptotically stable. This is the case indeed if the equilibrium point lies to the right of the maximum point of the prey isocline or the predator's diffusion coefficient d_P is not larger than the prey's d_N. The situation is different if the equilibrium point lies in the Allée-effect zone in such a way that the equilibrium is an asymptotically stable solution of the kinetic system. In order to have this situation it is sufficient that in addition to (2.5.22) the conditions $\Theta_2 > 0$ and $\Theta_1\Theta_2 < \Theta_3\Theta_4$ be fulfilled. This will be assumed in the sequel. If we start to increase the predator diffusion coefficient we arrive at a value such that one of the eigenvalues of the coefficient matrix of system (2.5.24) with λ_1 substituted will be zero, the other one negative, and for all other λ_j-s the systems (2.5.24) stay asymptotically stable. To the zero eigenvalue there belongs a constant solution col $[y_{111}, \ y_{211}]$ and to this a *spatially nonconstant stationary solution* of the boundary value problem (2.5.23):

$$\begin{bmatrix} v_1(x) \\ v_2(x) \end{bmatrix} = \begin{bmatrix} y_{111} \\ y_{211} \end{bmatrix} \cos(\pi x/l)$$

while the rest of the terms in the series (2.5.27) tend to zero as t tends to infinity. This spatially nonconstant time-independent solution is called a *pattern*. Its occurrence in the linearized problem has implications for the original nonlinear problem. With fairly deep mathematics of nonlinear functional analysis it has been proved that under generic conditions if

$$\eta\Theta_1\Theta_2/\lambda_1 > d_N \geq \eta\Theta_1\Theta_2/\lambda_2$$

and d_P is increased through the value

$$d_{P\text{crit}} = (\lambda_1 d_N \eta\Theta_3\Theta_4 + \det A)/(\lambda_1\eta\Theta_1\Theta_2 - \lambda_1^2 d_N)\,,$$

then the equilibrium $(\overline{N}, \overline{P})$ of the boundary value problem (2.5.19)–(2.5.20) undergoes a Turing bifurcation. This means that $(\overline{N}, \overline{P})$ is an asymptotically stable solution of the problem if d_P is below the critical value and it becomes unstable if d_P is above it. At the same time, for values of d_P near to the critical value a pattern, that is, a spatially nonconstant stationary solution, occurs (see Fig. 2.5.4).

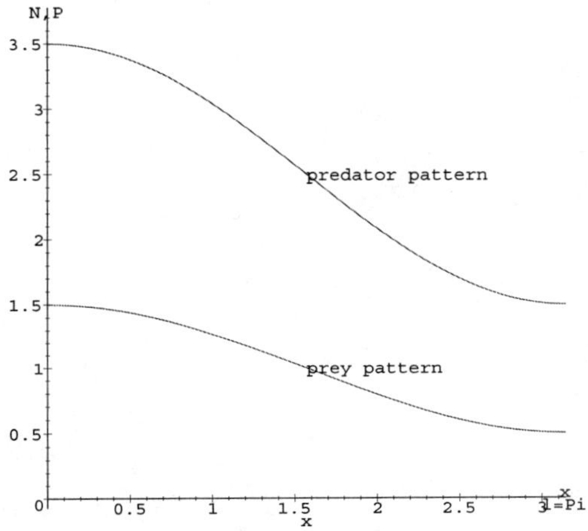

Figure 2.5.4: Approximate prey and predator patterns on the interval $(0, \pi)$ with $(\overline{N}, \overline{P}) = (1, 2.5)$. (MAPLE-V).

Chapter 3

EPIDEMICS

In this chapter mathematical models will be studied that describe the spread of epidemics in a population. Although great mathematicians such as Euler and D. Bernoulli have already tried to describe the dynamics of epidemics by mathematical methods, the modern mathematical theory probably received impetus when papers by Kermack and McKendrick (1927, 1932, 1933) were published. Such models, even the most primitive ones, may help to find those points where one may most successfully fight an epidemic or forecast how it will pass. We deal first with the classical and simplest, the so-called SIR (Susceptibles, Infectives, Removed) model, then we consider the case of sexually transmitted diseases and so-called SIS (Susceptibles, Infectives, Susceptibles) models. For sexually transmitted diseases we treat the problem of pair formation in the human population. Finally, the spread of epidemics through space will be studied. Important references in the field are Murray (1989) and Capasso (1993).

3.1 The Spread of Diseases and Susceptibles/Infectives/Removed Models

These models are simple but still may yield some insight into the dynamics of a contagious illness in a densely populated city, an army barrack, or a student dormitory. Basic assumptions for such a model are:

(i) the total population is constant; the epidemic does not have a recognizable influence on population numbers;

(ii) the population is "well stirred," meaning that every individual has an equal chance to meet any other member of the population; and

(iii) any person in the population who caught the disease either obtained immunity or died (from the point of view of the individual this difference is far from being irrelevant but for the model, *horribile dictu*, the effect is the same).

We denote the number of *susceptibles, infectives* and those who obtained immunity (or died) called *removed* at time t by $S(t)$, $I(t)$ and $R(t)$, respectively. We assume that susceptibles move into the group of infectives through infection and a decrease in their number in unit time is proportional to the number of encounters of a susceptible and an infective individual and this, in turn, is proportional to their respective numbers. Denoting the *infection rate* by $r > 0$, the differential equation governing the variations in the number of susceptibles is then $\dot{S} = -rSI$. The class of infectives is recruited from the susceptibles by incorporating those who leave the latter class, and it is decreased by recovery from the disease (or death). Denoting the *recovery rate* by $a > 0$, the differential equation for the infectives is $\dot{I} = rSI - aI$; finally, the recovered follow the equation $\dot{R} = aI$. Thus, we have arrived at a 3D system of differential equations

$$\dot{S} = -rSI, \qquad \dot{I} = rSI - aI, \qquad \dot{R} = aI. \tag{3.1.1}$$

This system satisfies requirement (i), where the total population has constant size because by adding the three equations we get $(S + I + R)^{\cdot} = 0$, that is, the sum of the sizes of the three classes $N := S + I + R$ is constant. It is to be noted that in this model there is no latent period for the illness; a susceptible person who has contracted the disease becomes infective immediately. If incubation is short this abstraction may be accepted. The initial conditions attached to the system are $S_0 = S(0) > 0, I_0 = I(0) > 0, 0 = R(0)$ with $S_0 + I_0 = N$. Usually, one may assume that the disease starts with a small number of infectives, that is, I_0 is small with respect to S_0. The basic problem is whether or not the number of infectives will increase at the beginning. We say that we have an *epidemic* on hand if at the outset the number of infectives is increasing, which means that people get infected faster than they recover. It is easy to give a condition for that because from the second equation

$$\dot{I}(0) = I_0 (rS_0 - a) > \text{ respectively } < 0,$$

according to

$$S_0 > \text{ respectively } < a/r.$$

From the first of Eqs. (3.1.1) it is clear that $\dot{S}$ is always negative, so that $S(t) < S_0$ for $t > 0$. Thus, if $S_0 < a/r$ then $\dot{I}(t) = I(t)(rS(t) - a) < I(t)(rS_0 - a) < 0$ for $t \geq 0$; this means that if at the very beginning the number of infectives was decreasing this will remain. On the other hand, if the condition for an epidemic holds at the beginning then the number of infectives will increase for at least some time. The threshold parameter a/r is called the *relative recovery rate*, which is the percentage of those recovered in unit time divided by the percentage of those infected by a single infective in unit time. We shall give here an intuitive interpretation of the condition of the outbreak of an epidemic. First it is to be noted that the reciprocal $1/a$ of the recovery rate can be interpreted as the average infectious period of an infective or the average time needed for recovery. This can be seen from the third equation–if we have just one infective at time t, that is, $I(t) = 1$ then during the time interval $1/a$ the number of recovered will change by $R(t+1/a) - R(t) \approx \dot{R}(t)\, 1/a = I(t) = 1$, meaning that the single infective recovers. From the first equation during the time interval $1/a$ the number of those getting the infection is $S(t+1/a) - S(t) \approx \dot{S}(t)\, 1/a = -(r/a)\, S(t)$. If this number is < -1, one infective passes on the disease during his infectious period to more than one person, thus if $(r/a)\, S(t) > 1$, implying that

$$S_0 r/a > 1, \tag{3.1.2}$$

then more persons get ill in unit time than recover; thus an epidemic breaks out. The ratio r/a, called the *contact rate*, is the infection rate multiplied by the average infectious period. The contact rate multiplied by the number of susceptibles gives the number of those infected by one infective during the infectious period of the latter. We determine now the projections of the trajectories of system (3.1.1) on the S, I plane. For this we divide the second equation by the first one to obtain the differential equation of the trajectories, $dI/dS = -1 + a/(rS)$. It is easy to integrate yielding, as the equation of the trajectories,

$$I = \ln(S^{a/r}) - S + c \tag{3.1.3}$$

where $c = I_0 + S_0 - (a/r)\ln S_0$. It is easy to see that for all these trajectories $\max I(S) = I(a/r)$, provided that a/r is $< S_0$ (see Fig. 3.1.1 produced by MAPLE).

Thus, we see that if the number of susceptibles is greater than this threshold value at the start then the number of infectives will rise at the beginning until the number of susceptibles decreases enough to reach this value. If the number of susceptibles is less than the threshold value at the beginning, then the number of infectives is going down right away. By dividing the first equation by the third one, it is easy to obtain S as a function of R; expressing I in the third equation by S and R and substituting $S(R) = S_0 \exp(-Ra/r)$ obtains a first-order scalar differential equation for R. Solving the latter by MAPLE, we obtain the number of removed as a function of time (Fig. 3.1.2, thick line). Finally, the number of removed per unit time (the mortality in case of a lethal disease), that is, the derivative of R with respect to time is shown in Fig. 3.1.2 (dotted line).

As the figures show even this simple model is able to help us in forecasting the number of all those who will catch the disease during the epidemic, the time when the number of infectives will be maximal, the time when the mortality will be maximal, and the time when the epidemic may be considered to be over etc., provided that we know the infection and recovery rates and the initial data. Naturally, refinements can be made in several directions–to take into consideration the incubation period, count the "carriers" who are not ill themselves but spread the disease, and consider the age structure etc.

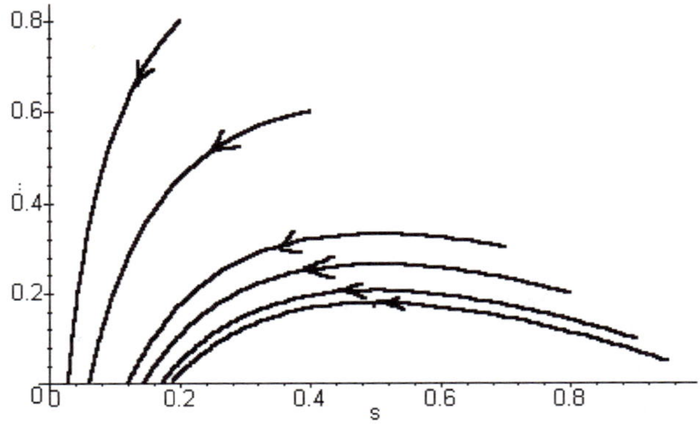

Figure 3.1.1: The spread of disease using the SIR model (3.1.1) with total population 1, infection rate $r = 2$, recovery rate $a = 1$, and relative recovery rate $a/r = 0.5$; the number of infectives versus the number of susceptibles in the first 10 time units of the outbreak (MAPLE).

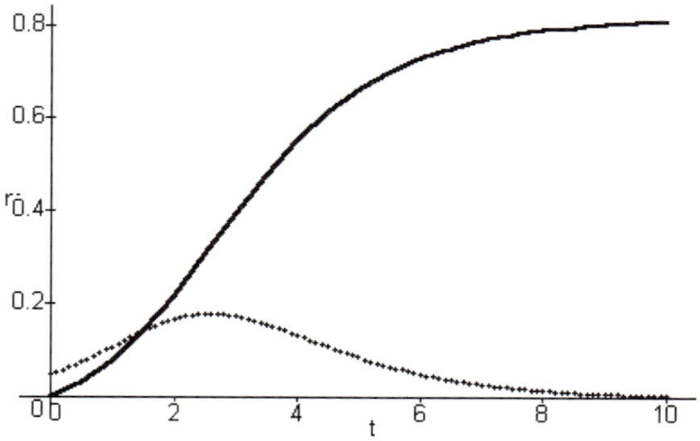

Figure 3.1.2: Thick line is the number of removed in the first 10 units of time and the dotted line is the mortality (in case of a lethal disease) in the first 10 units of time $I(0) = 0.05$ (MAPLE).

3.2 Sexually Transmitted Diseases

Venereal diseases differ from other epidemics prevalent in the human population in that the population is divided into two groups, males and females, and the disease is normally transmitted only from a member of one of the groups to a member of the *other* group, that is, males transmit the disease to females and vice versa. Further, because no immunity is conferred by going through the disease, if a person passes from the susceptible group to the infective one then following recovery from the disease he/she becomes susceptible again. Models without immunity are called SIS models. Naturally, if one considers AIDS lethal then it does not fall into this category. The model we handle in this Section considers a disease such as gonorrhea. Many mathematical models of this particular illness were presented by Hethcote and Yorke (1984). We present the simplest one.

We assume that the total population of sexually active males and females is constant. The number of susceptible males and females at time t is denoted by $S_1(t)$ and $S_2(t)$, respectively, and the number of infective males and females by $I_1(t)$ and $I_2(t)$. By assumption $S_1(t) + I_1(t) = N_1$, $S_2(t) + I_2(t) = N_2$, where N_1 and N_2 are constants. The number of susceptible males decreases in unit time by those who get infected by infective females and vice versa, the number of males getting the infection in unit time is supposed to be proportional to the number of susceptibles and to the number of infective females. The number of susceptible males/females increases by those who recover from the disease–the number of those recovered in unit time is proportional to the number of infective males/females. The number of infectives increases in unit time by those who get the infection and decreases by those who recover. This way we arrive at the following four-dimensional (4D) system of differential equations:

$$\begin{aligned} \dot{S}_1 &= -r_1 S_1 I_2 + a_1 I_1 , \qquad & \dot{S}_2 = -r_2 S_2 I_1 + a_2 I_2 \\ \dot{I}_1 &= r_1 S_1 I_2 - a_1 I_1 , \qquad & \dot{I}_2 = r_2 S_2 I_1 - a_2 I_2 , \end{aligned} \tag{3.2.1}$$

where r_1, a_1 and r_2, a_2 are the infection and recovery rates of males, and females, respectively (cf. system (3.1.1)). Taking into account that the sum of susceptible and infective males/females is constant, this system can be reduced to a 2D one:

$$\dot{I}_1 = r_1(N_1 - I_1)I_2 - a_1 I_1 , \qquad \dot{I}_2 = r_2(N_2 - I_2)I_1 - a_2 I_2 . \tag{3.2.2}$$

This system is such that if $I_1 = 0$ then $\dot{I}_1$ is positive and, similarly, if $I_2 = 0$ then $\dot{I}_2$ is positive. This means that the positive quadrant of the I_1, I_2 plane is positively invariant–no trajectory may leave the positive quadrant. The system has two equilibria: $(0, 0)$ and

$$\begin{aligned} (\bar{I}_1, \bar{I}_2) = (&(N_1 N_2 - a_1 a_2 / (r_1 r_2)) / (N_2 + a_1/r_1) , \\ &(N_1 N_2 - a_1 a_2 / (r_1 r_2)) / (N_1 + a_2/r_2)) , \end{aligned}$$

with the last one in the positive quadrant iff $N_1 N_2 - a_1 a_2 / (r_1 r_2) > 0$, or

$$(N_1 r_1 / a_2)(N_2 r_2 / a_1) > 1 . \tag{3.2.3}$$

The last inequality has a clear intuitive meaning. The first factor on the left-hand side is the number of males multiplied by the contact rate with respect to males, that is the infection rate of males multiplied by the infectious period of an infective *female.* Thus, it gives the number of males infected by an infective female during her infective period. (To be sure, here N_1 ought to be replaced by the number of *susceptible* males but if the disease is not too widely spread the difference is slight.) The second factor has an analogous meaning for females. Hence, we see that the condition for the existence of an *endemic equilibrium* $(\bar{I}_1, \bar{I}_2)$ in the positive quadrant is that on the average one infective shall infect more than one person during his/her infective period. Naturally, the main question is whether the epidemic free state $(0,0)$ or the endemic state $(\bar{I}_1, \bar{I}_2)$ is stable. In order to be able to tell this we linearize the system at the two equilibria and apply the Routh-Hurwitz criterion (see Theorem A1.1.2). At $(0,0)$ we leave this to the reader as an exercise. It turns out by an easy calculation that if (3.2.3) is reversed then the epidemic free state is asymptotically stable, if (3.2.3) holds then it is unstable (as a matter of fact, a saddle point). The calculation for the endemic state is more tiresome. From now on (3.2.3) is assumed. The characteristic polynomial is

$$\begin{vmatrix} -a_1 - r_1 \frac{N_1N_2 - a_1a_2/(r_1r_2)}{N_1 + a_2/r_2} - \lambda & r_1\left(N_1 - \frac{N_1N_2 - a_1a_2/(r_1r_2)}{N_2 + a_1/r_1}\right) \\ r_2\left(N_2 - \frac{N_1N_2 - a_1a_2/(r_1r_2)}{N_1 + a_2/r_2}\right) & -a_2 - r_2\frac{N_1N_2 - a_1a_2/(r_1r_2)}{N_2 + a_1/r_1} - \lambda \end{vmatrix}$$
$$= \lambda^2 + \lambda\left(a_1 + a_2 + r_1\frac{N_1N_2 - a_1a_2/(r_1r_2)}{N_1 + a_2/r_2} + r_2\frac{N_1N_2 - a_1a_2/(r_1r_2)}{N_2 + a_1/r_1}\right)$$
$$+ \frac{(N_1N_2 - a_1a_2/(r_1r_2))(a_1r_2N_1 + a_2r_1N_2 + r_1r_2N_1N_2 + a_1a_2)}{(N_2 + a_1/r_1)(N_1 + a_2/r_2)}.$$

By (3.2.3) both the coefficient of λ and the "constant term" are positive, hence with Theorem A1.1.2 the endemic equilibrium (if it exists in the interior of the positive quadrant) is asymptotically stable.

Condition (3.2.3) of an asymptotically stable endemic equilibrium shows clearly how the decrease of the average infectious period $1/a_1$ and $1/a_2$, respectively, and/or the infection rate r_1 and r_2, respectively, may destabilize the endemic state and lead to a disease-free state. If, for instance, we assume that the average infectious period of a male is 1.5 months and of a female is 3.5 months, respectively, the respective infection rates are $r_1 = 1.4 \cdot 10^{-8}$ per month and $r_2 = 3.8 \cdot 10^{-8}$ per month, and both the total sexually active male and female population are equal to $20 \cdot 10^6$, then the left-hand side of (3.2.3) is $1.127 > 1$ and the stable endemic equilibrium is $(\bar{I}_1, \bar{I}_2) = (0.515 \cdot 10^6, 1.27 \cdot 10^6)$. If the average infectious period of women could be brought down to 2.5 months then the left-hand side of (3.2.3) would become $0.805 < 1$, the endemic equilibrium would disappear, and the disease-free state would become stable.

3.3 A Model of Pair Formation

Early models on the dynamics of sexually transmitted diseases including the one treated in the previous Section have the disadvantage of being based on the assumption of "well stirredness" of the population, that is, it is assumed that each member of the population has an equal chance to meet any other member and mating is completely random. This assumption is clearly false for the human population, although there might be small highly promiscuous subgroups in which this assumption may be a not completely incorrect approximation of prevailing behavior. To better approximate reality one has to study how pairs are formed, how a sexual partnership emerges and how it ceases to exist. If a female and a male form a pair and they do not carry the disease then they can be considered immune until one of them does not have contact with another partner. The duration of these partnerships and the time interval between two partnerships of a given person are to be taken into account. If the problem of *pair formation* has been settled then one may build a model based on it to describe the spread of a sexually transmitted disease. Several models of pair formation can be found in the literature, with some of them dividing up the population to several less or highly promiscuous subgroups in which pair formation is governed by different laws. We present here a fairly simple one due to Dietz and Hadeler (1988) (see also the references therein) which is highly instructive, and then we shall describe how an epidemic model can be built upon it.

Denote the density of single females and males by x and y, respectively, suppose that their densities are increasing by constant rates κ_x and κ_y, respectively, due to the aging of younger generations into a sexually active population (we do not consider here dependence of the birth rate on the density of the population) and that these densities decrease due to deaths proportional to the numbers with mortalities μ_x and μ_y, respectively. Denote the density of pairs formed by a female and a male by p. At this point we must decide what we consider a pair–how we define a pair. If we want to use the model for describing the spread of sexually transmitted diseases then the social or religious aspects of pair formation have to be disregarded. We say that a *pair* is formed when a female and a male have sexual contact with each other the first time and this pair ceases to exist the first time one of the members has sexual contact outside the pair. Denote the *divorce rate* (considered to be a positive constant) by σ. The number of single females and males increases by one when a pair separates, and the number of single males and females, respectively, increases by one when the female or male member of a pair dies. Pairs are recruited from the single population. The rate of pair formation is a function of x and y; it is sometimes called the *marriage function* and will be denoted by $\varphi(x, y)$. It has to satisfy certain natural conditions: (i) it is defined for nonnegative values of x and y and must be zero when either the females or the males are absent, $\varphi(0, y) = \varphi(x, 0) = 0$; (ii) it is increasing (or, rather, not decreasing) if the number of single females or males is increasing, $\varphi'_x, \varphi'_y \geq 0$; (iii) if the densities of both single females and males are increasing $\alpha > 0$ times the rate of pair formation also increases α times; this means that the function φ is homogeneous of degree one: $\varphi(\alpha x, \alpha y) = \alpha\varphi(x, y)$; and (iv) φ is symmetric, $\varphi(x, y) = \varphi(y, x)$. With these assumptions and notations we arrive at the following system of differential equations:

$$\begin{aligned} \dot{x} &= \kappa_x - \mu_x x + (\mu_y + \sigma) p - \varphi(x, y) \\ \dot{y} &= \kappa_y - \mu_y y + (\mu_x + \sigma) p - \varphi(x, y) \\ \dot{p} &= -(\mu_x + \mu_y + \sigma) p + \varphi(x, y) \ . \end{aligned} \tag{3.3.1}$$

Here we have also tacitly assumed that the mortality of singles is equal to the mortality of those in pairs, even though sociological data suggest that people in permanent partnership live longer than singles. There are several possible ways to choose a marriage function that satisfies conditions (i)-(iv); here we suppose that *the rate of pair formation depends linearly on the density of that sex that is in minority.* The choice representing this assumption is

$$\varphi(x, y) := \rho \min(x, y) = \begin{cases} \rho x, & \text{if} \quad (x, y) \in K_y \\ \rho y, & \text{if} \quad (x, y) \in K_x \ , \end{cases} \tag{3.3.2}$$

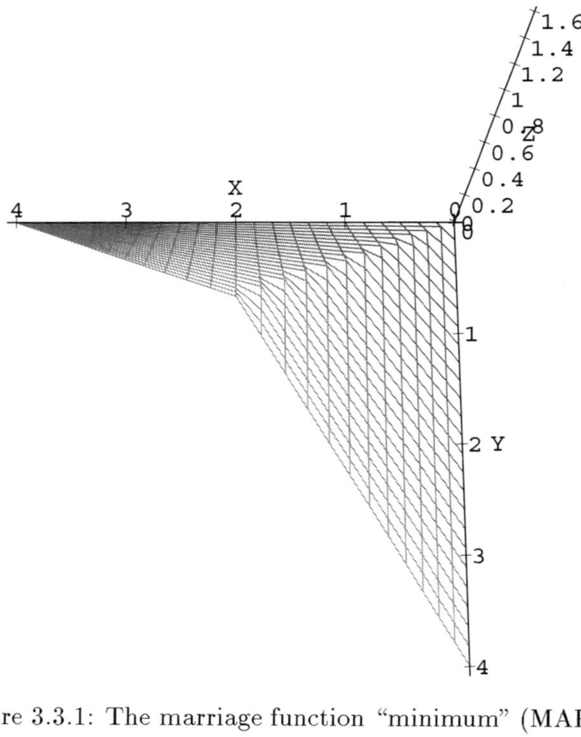

Figure 3.3.1: The marriage function "minimum" (MAPLE).

where $K_y = \{(x, y) : 0 \leq x \leq y\}$, $K_x = \{(x, y) : 0 \leq y \leq x\}$, and ρ is a positive constant. The graph of this function, which looks like the part of a roof, can be seen in Fig. 3.3.1 (produced by MAPLE). It is made up by two planes intersecting over the line $y = x$ of the plane x, y.

In most societies the number of females is increasing faster than the number of males and the mortality of women is lower than that of the men. Therefore we assume in the sequel that

$$\kappa_x \geq \kappa_y \text{ and } \mu_x \leq \mu_y \ . \tag{3.3.3}$$

This way we arrive at the so-called *male dominance model.* Under condition (3.3.3) the set K_x representing female majority is positively invariant, that is, no trajectory may leave K_x because on the boundary $x = y$ we have

$$\begin{aligned} (x-y)^{\cdot} &= \kappa_x - \kappa_y - \mu_x x + \mu_y y + (\mu_y - \mu_x)p \\ &\geq \kappa_x - \kappa_y - \mu_x (x-y) + (\mu_y - \mu_x)p \\ &= \kappa_x - \kappa_y + (\mu_y - \mu_x)p \geq 0, \end{aligned}$$

and thus, if the number of females was greater than the number of males, this stays so, and on the boundary $y = 0$ we have from the second equation of system (3.3.1): $\dot{y} = \kappa_y + (\mu_x + \sigma)\, p > 0$. In this case we may replace the function φ by ρy in the system over K_x and we have to deal with the linear system

$$\begin{aligned} \dot{x} &= \kappa_x - \mu_x x - \rho y + (\mu_y + \sigma)\, p \\ \dot{y} &= \kappa_y - (\mu_y + \rho) y + (\mu_x + \sigma)\, p \\ \dot{p} &= \rho y - (\mu_x + \mu_y + \sigma)\, p \, , \end{aligned} \tag{3.3.4}$$

$(x, y) \in K_x, \quad p \geq 0$. This model can be explicitly calculated. It has a unique equilibrium:

$$\begin{aligned} &(\bar{x}, \bar{y}, \bar{p}) \\ &= \left(\frac{\kappa_x}{\mu_x} - \frac{\kappa_y}{\mu_y} \frac{\rho}{\mu_x + \mu_y + \sigma + \rho}, \frac{\kappa_y}{\mu_y} \frac{\mu_x + \mu_y + \sigma}{\mu_x + \mu_y + \sigma + \rho}, \frac{\kappa_y}{\mu_y} \frac{\rho}{\mu_x + \mu_y + \sigma + \rho} \right) . \end{aligned}$$

It is easy to see that because of (3.3.3) $(\bar{x}, \bar{y}) \in K_x$. We note that in equilibrium the number of females and the number of males is $\bar{f} = \bar{x} + \bar{p} = \kappa_x / \mu_x$ and $\bar{m} = \bar{y} + \bar{p} = \kappa_y / \mu_y < \bar{f}$, respectively. A simple calculation yields the characteristic polynomial of the coefficient matrix of system (3.3.4), it is

$$\begin{aligned} &\lambda^3 + \lambda^2 (2\mu_x + 2\mu_y + \sigma + \rho) \\ &+ \lambda \left(\mu_x (\mu_x + 2\mu_y + \sigma + \rho) + \mu_y (\mu_x + \mu_y + \sigma + \rho) \right) \\ &+ \mu_x \mu_y (\mu_x + \mu_y + \sigma + \rho) \, . \end{aligned}$$

All the coefficients are positive and one may check easily that condition (A1.1.1) is also satisfied, so that the equilibrium $(\bar{x}, \bar{y}, \bar{p})$ is globally asymptotically stable in $K_x \times \mathbf{R}_+$.

From the point of view of the spread of disease the average length of a partnership or the mean number of partners during a lifetime is crucial. In order to estimate this we determine first the mean lifetime of a female and a male. In complete analogy to how the average infectious period has been shown to be the reciprocal of the recovery rate (Section 3.1 preceding (3.1.2)) from system (3.3.1) we obtain that the mean active lifetime of a female and a male is $1/\mu_x$ and $1/\mu_y$, respectively. For instance, if there are no pairs, males are not recruited, and at time t there is just 1 male present, then at time $t + 1/\mu_y$ the number of males present will be $y\,(t + 1/\mu_y) \approx y\,(t) + \dot{y}\,(t)\,(1/\mu_y) = 1 - \mu_y \cdot 1 \cdot (1/\mu_y) = 0$. Similarly, the average duration of a partnership is $1/\,(\mu_x + \mu_y + \sigma)$. If no pairs are present at time t, and there is just 1 male, then from the third equation of system (3.3.4) $p\,(t + 1/\rho) - p\,(t) = p\,(t + 1/\rho) \approx \dot{p}\,(t)\,(1/\rho) = \rho \cdot 1 \cdot (1/\rho) = 1$, and thus, the average time needed for the formation of a pair (for finding a partner) is $1/\rho$. As a consequence, the time consumed by the search for a partner plus the time spent in this partnership is $1/\rho + 1/\,(\mu_x + \mu_y + \sigma) = (\mu_x + \mu_y + \sigma + \rho)\,/\,(\rho\,(\mu_x + \mu_y + \sigma))$. The product of this duration and the average number of partners of a male during lifetime N_y must be equal to the mean active lifetime of a male:

$$N_y\,(\mu_x + \mu_y + \sigma + \rho)\,/\,(\rho\,(\mu_x + \mu_y + \sigma)) = 1/\mu_y\,.$$

Hence, the average number of partners of a male during lifetime is

$$N_y = \frac{\rho}{\mu_y} \frac{\mu_x + \mu_y + \sigma}{\mu_x + \mu_y + \sigma + \rho}\,.$$

It is reasonable to suppose that the ratio of the average number of partners of a female and that of a male is equal to the ratio of their respective lifetimes divided by the ratio of their numbers in equilibrium:

$$\frac{N_x}{N_y} = \frac{1/\mu_x}{1/\mu_y} \frac{\bar{m}}{\bar{f}} = \frac{\kappa_y}{\kappa_x} \leq 1$$

by (3.3.3).

Dietz and Hadeler (1988) set up the model of pair formation and then built a model for the spread of the disease. Due to lack of space we can not present this model in detail here but we shall describe its main features. The population is divided into 8 groups: noninfected and infected females and males and pairs in which both partners are noninfected, both are infected, only the male or only the female, respectively, is infected. This way a system of eight dimensions is constructed in a fairly straightforward way. If there is no infection in the population the system reduces to system (3.3.1) or (3.3.4). A condition can be given for the stability of the disease-free equilibrium of the latter system given in the preceeding from the point of view of the eight-dimensional (8D) system. If one assumes that the rates of increase, death, infection, recovery etc. do not depend on the sex one obtains a simpler five-dimensional (5D) system of differential equations. Even in this lower dimensional model one may get explicit results by analytical methods only if recovery is excluded (which is the case now if one tries to apply the results to HIV). Under this assumption a threshold condition can be given that implies the existence of an endemic equilibrium. With the data assumed by the authors 4 years of average duration of partnership is the threshold below which the endemic equilibrium persists. Four years of duration corresponds to an average of 12 partners per 50 years of active lifetime.

3.4 The Spread of Epidemics in Space

In the classical models of epidemics the "well stirredness" assumption made at the beginning of Section 3.1 plays a crucial role. In the previous Section, as long as sexually transmitted diseases were involved, we got rid of the "random mating" part of it by taking into account the dynamics of pair formation in the human population. Up to this point, however, the population was considered to be concentrated in one point and it was not taken into account that, in fact, the population has a spatial distribution on a continent, in a country, or even in a large town. In previous centuries diseases such as plague, cholera, or influenza swept over continents like a wave spreading from one place to the neighboring one and so on. To be sure, in the twenty-first century, distances probably are not as important as they were because a few passengers on a plane from, for example, Hong Kong who are carrying influenza may bring it into an American or European capital before it even reaches Shanghai. However, it remains undeniable that some diseases spread from place to place and people who live far away from a disease source may have a better chance of avoiding epidemics than those who live near the nucleus. Therefore, in this Section a model will be treated in which the spatial distribution of the population will be taken into account. This model was used by Murray (1989) to describe the "Black Death," the bubonic plague that swept through Europe from 1347 to 1350 and killed about one-quarter of the population (see Langer, 1964). We treat the model somewhat differently here but recommend reading of the vivid description of the case in the literature quoted.

We denote the *areal density of susceptibles and infectives* at time t and at place x by $S(t,x)$ and $I(t,x)$, respectively, and the *infection rate* and the *mortality of infectives* by $r > 0$ and $a > 0$, respectively. A SIR model is to be built but the equation for the removed is not written out and it is assumed that susceptibles and infectives move around, following Fick's diffusion law (see Appendix 3.2) with a diffusion rate $D > 0$. The following system of partial differential equations describes the dynamics:

$$\frac{\partial S}{\partial t} = -rSI + D\Delta S\,, \qquad \frac{\partial I}{\partial t} = rSI - aI + D\Delta I\,, \tag{3.4.1}$$

where Δ is the Laplace delta-if the space is 2D (the surface of a part of the earth considered to be a plane) and x and y are Cartesian orthogonal coordinates $\partial^2/\partial x^2 + \partial^2/\partial y^2$, and if the space is 1D (straight line in a certain direction) just the second derivative with respect to the spatial variable $\partial^2/\partial x^2$. If there is no diffusion, $D = 0$ then we get back system (3.1.1) without the equation for the removed. Although the problem could be treated in the realistic two spatial dimension setting, in order to simplify the mathematics we are to treat it by assuming a 1D space. This is surely an abstraction but it still may give an insight into the dynamics of the propagation of disease if we suppose that the epidemic propagates from the nucleus uniformly in every direction. Further, we suppose that the domain where the disease propagates is infinite without boundary. Although the problem could be treated also by assuming, say, no flux boundary conditions, the complications at the boundary would increase the mathematical difficulties without much gain. It is assumed that the diffusion rates of the susceptibles and the infectives are the same. One may say that this is not too realistic because the sick do not move around. However, during the incubation period there is no difference in behavior among those who carry the disease and those who do not. During large plague epidemics in Europe there was massive emigration from the large towns, which served to accelerate propagation (the most famous literary evidence of this is, perhaps, the setting of Boccaccio's *Decameron*, in which a company of gentlemen and ladies who fled the plague for a country house tell each other spicy stories). The plague was carried also by rats, and no one knows now what was the diffusion rate of healthy and sick rats. In model (3.4.1) the susceptible population is considered to be constant if infectives are not present and on a far away boundary there is no in- and outflow of people, that is, no birth and death process apart from the epidemics is taken into account. We suppose that those who get the disease die; in case of the plague, indeed, 80–90% of those who fell ill did not recover. Under all these conditions we are to treat system (3.4.1) in one spatial dimension,

$$\frac{\partial S}{\partial t} = -rSI + D\frac{\partial^2 S}{\partial x^2}, \qquad \frac{\partial I}{\partial t} = rSI - aI + D\frac{\partial^2 I}{\partial x^2}. \tag{3.4.2}$$

We may simplify the equations by changing the scales introducing the new variables

$$h = \frac{S}{S_0}, \qquad v = \frac{I}{S_0}, \qquad \tau = rS_0 t, \qquad y = \left(\frac{rS_0}{D}\right)^{1/2} x, \tag{3.4.3}$$

where S_0 is the initial value of susceptible density. A simple calculation yields the system in the new coordinates:

$$\frac{\partial h}{\partial \tau} = -hv + \frac{\partial^2 h}{\partial y^2}, \qquad \frac{\partial v}{\partial \tau} = hv - bv + \frac{\partial^2 v}{\partial y^2}, \tag{3.4.4}$$

where $b = a/(rS_0)$, its reciprocal rS_0/a is the contact rate multiplied by the density of susceptibles, giving the density of those infected by a unit density of infectives (cf. Section 3.1 and especially the discussion on condition (3.1.2)). The equilibria of system (3.4.4) are $(h, v) = (h_0, 0)$ with arbitrary $h_0 > 0$; however, $(h, v) = (1, 0)$ corresponds to the initial value $(S, I) = (S_0, 0)$. Rather than treating system (3.4.4) in a general way, we try to find only those solutions that describe the spread of the disease in the form of a wave, that is *traveling wave* solutions. These are the solutions that depend on τ and y only through the expression $z = y - c\tau$ with some constant $c > 0$. In the "space-time" plane y, τ along the straight lines $y - c\tau = z$ with a constant z the state of the system is the same and the values of h, v are constant. If we pick two points on one of these straight lines (y_1, τ_1) and (y_2, τ_2), that is, $y_1 - c\tau_1 = z = y_2 - c\tau_2$ and $\tau_1 < \tau_2$, for example, then this means that if the system was in a certain state at moment τ_1 at place y_1 then it will be in the same state at moment τ_2 at place y_2. This obviously means that the state of the system is propagating along these parallel lines with *velocity* $c = (y_2 - y_1) / (\tau_2 - \tau_1)$. If we are interested in solutions of the form $h(z) = h(y - c\tau)$, $v(z) = v(y - c\tau)$ then the system reduces to the system of ordinary differential equations:

$$\frac{d^2h}{dz^2} + c\frac{dh}{dz} - hv = 0\,, \qquad \frac{d^2v}{dz^2} + c\frac{dv}{dz} + (h - b)\,v = 0\,. \tag{3.4.5}$$

The equilibria of this system are also $(h_0, 0)$; in particular, $(1, 0)$ interests us. The question is, what are the solutions doing as z tends to infinity and to minus infinity? If place y is fixed then *as time τ tends to infinity the variable z tends to minus infinity and vice versa and as τ tends to minus infinity z tends to plus infinity.* System (3.4.5) will be linearized at $(1, 0)$ and the eigenvalues will be determined. First we put the system into Cauchy normal form introducing the new phase variables,

$$x_1 = h, \qquad x_2 = \dot{h}, \qquad x_3 = v, \qquad x_4 = \dot{v}\,.$$

The derivative with respect to z is denoted with an overdot and we obtain

$$\begin{aligned} \dot{x}_1 &= x_2\,, \qquad \dot{x}_2 = -cx_2 + x_1x_3\,, \\ \dot{x}_3 &= x_4\,, \qquad \dot{x}_4 = -cx_4 + (b - x_1)\,x_3\,. \end{aligned}$$

Linearizing at (1,0) the characteristic polynomial turns out to be

$$p(\lambda) = \lambda\,(c + \lambda)\left(\lambda^2 + c\lambda + 1 - b\right)\,.$$

The eigenvalues are 0, $-c$, $(1/2)\left(-c \pm \sqrt{c^2 + 4\,(b - 1)}\right)$. That one of the eigenvalues is zero is no wonder because the equilibria fill in the axis h, which is the center manifold of dimension one of each (see Appendix 2.3). The third and fourth eigenvalues have negative real parts iff a number $< c^2$ stands below the square root, that is, if $b < 1$. Assuming this we must also suppose that the expression below the square root is nonnegative, that is, $c \geq 2\sqrt{1 - b}$. Otherwise the solutions of the linearized system and with them the solutions of the original nonlinear system would oscillate around (1,0) with x_1 and x_3 assuming negative values that have no meaning. Under these assumptions the solutions tend towards the equilibria on the axis h as z tends to infinity, that is, time tends to minus infinity. This means that if the system is perturbed out of the equilibrium (1,0) by a small positive initial value of infectives v then the solutions tend away from the equilibrium and a traveling wave of epidemics starts to propagate as *time is increasing*. Summing up what has been established already here, a traveling wave of epidemics sweeps through a region if $a/(rS_0) < 1$ or

$$\frac{r}{a}S_0 > 1\,, \qquad (3.4.6)$$

that is, if the density of those infected by a unit density of infectives during their infectious lifetime is greater than one. The minimal velocity of the wave sweeping through the population in the region is $c = 2\sqrt{1-b}$ in the transformed coordinates, or as $x/t = (rS_0D)^{1/2}\,y/\tau$,

$$c_{\text{real}} = 2\sqrt{rS_0D - aD} \qquad (3.4.7)$$

in the original spatial and time scale.

We try to fit this model to the "Black Death" epidemic of 1347-1350 that swept through Europe, starting from the port of Genoa (Genova earlier), on the western coast of Italy and reaching Russia, the Baltic, and Sweden in only 3 years. Most of the data used here are taken from Murray (1989). Eighty to 90% of those who contracted the plague died on average within 12 days, so that the infectious period including the incubation is $1/a = 12\ days = 0.033\ years$, meaning that $a = 30\ year^{-1}$. The infection rate is estimated at $r = 0.4\ mile^2/year$. The population density of Europe was at that time estimated to be $50/mile^2$; however, because not only men but also rats carry the fleas that carry the disease one has to increase this number considerably, so that we double this figure to $S_0 = 100/mile^2$. This yields $b = 0.75$. From these data we obtain for the minimal transformed velocity of the epidemic wave $c = 1$. By Langer (1964) the average real velocity was $c_{\text{real}} = 400\ miles/year$. From Eq. (3.4.7) we may calculate the diffusion rate as $D = 4000\ mile^2/year$.

We have solved system (3.4.5) by PHASER using these data with two sets of initial values near $(h, v) = (1, 0)$ and small initial derivative values. The result is shown in Fig. 3.4.1. In plane h, v the projection of the trajectory shows how the density of susceptibles falls from 1.05 or 0.96, respectively, to 0.75 and 0.65, respectively, as the epidemic sweeps over the land. This corresponds to the estimate mentioned at the beginning of this Section that ≈ 25% of the population died in the epidemic. Then we show the graph of the susceptibles and infectives as functions of z. The horizontal axis z directed to the right is at the same time the axis y of the spatial coordinate at a fixed moment t; far to the right-hand side the population is not yet affected and is at its original level and far to the left-hand side the epidemic had already swept through and the population declined. The horizontal axis *directed to the left* corresponds to the time axis τ at a fixed place y; far to the right, there was no epidemic yet and so population density remained at the original level; over time, that is, toward the left, the effect of the epidemic began to be felt and the population began to decrease. One may look similarly at the graph of the infectives. At the height of the epidemic the maximal density of infectives was 0.034, and 0.022, respectively, meaning that ≈ 3% of the population was infected at the same time.

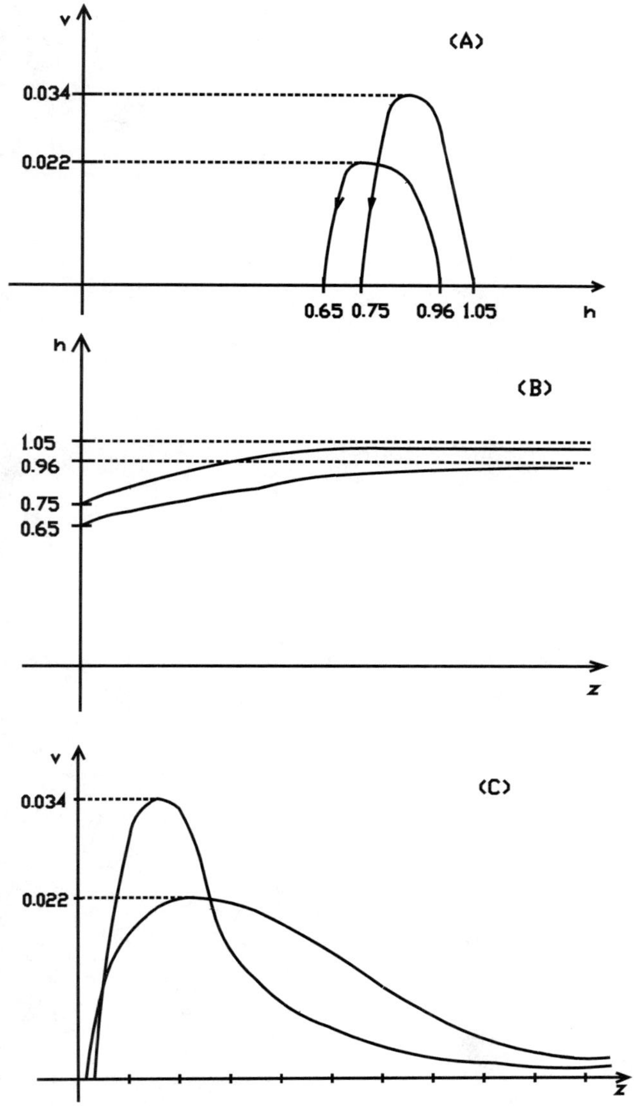

Figure 3.4.1: The traveling wave of an epidemic: system (3.4.5) $c = 1, b = 0.75$. (A) The trajectories in the plane h, v of susceptibles and infectives starting near $(1.05, 0)$ or $(0.96, 0)$ and ending at $(0.75, 0)$ or $(0.65, 0)$, respectively. (B) the graphs of the functions $h(z)$ of the two solutions; and (C) the graphs of the functions $v(z)$ of the two solutions; $z \sim y \sim -\tau$ (PHASER).

Chapter 4

EVOLUTION AND POPULATION GENETICS

In the first two Sections we shall deduce the mathematical formalism that expresses the basic principles governing the evolution of a single sexually reproducing species. In the third Section the current theory of the formation of macromolecules that are able to reproduce is presented in a concise form and in the fourth a glimpse of the vast and mostly undiscovered field dealing with the functioning of the immune system is provided. This chapter probably requires the most background in biology. It is impossible to insert a text of genetics here; we shall restrict ourselves to what is absolutely necessary for an understanding of the basic mathematical models treated.

4.1 The Phase Space of Genotypes and the Hardy-Weinberg Law

Each individual of our sexually reproducing species develops from a single *diploid* cell that is formed in the process of reproduction via the fusion of two *haploid gametes*, with one of these, the egg cell, coming from the mother, the other one, the sperm cell, from the father. These gametes carry genetic information to be passed from parents to offspring in the form of *genes* that are attached to specific chromosomal sites (*chromosomes*) within the gametes. The chromosomal site occupied by a certain type of gene with a certain function is called a *locus*. Normally there is a finite number of different genes that may occupy the same locus; these, called *alleles*, belong to a given locus. The alleles of a certain locus are responsible for the same heritable trait, for example, the color of the eyes, the blood group, or body size. We denote the allele set of a locus by $A_1, A_2, \ldots, A_n$. Which one sits on the given locus determines the *genome-type* of the gamete with respect to this locus. The chromosomes of the gamete are "chains" of such loci; the genome-type of the gamete is determined by the combination of alleles on all the loci on all chromosomes.

The chromosomes of the egg cell and those of the sperm cell are equal in number and they form homologous pairs after mating. The fusion of the haploid egg and the haploid sperm cell results in a *diploid zygote*, which therefore carries exactly two alleles on each locus. The *genotype of the zygote* with respect to a certain locus is determined by the two alleles that occupy the locus. On locus A, for example, there may sit allele A_i coming from the mother and allele A_j derived from the father, the genotype of the zygote with respect to locus A being A_iA_j. The genotype A_iA_j determines, in turn, the *phenotype* of the individual connected to locus A (the color of the iris, etc.) Genotypes A_iA_j and A_jA_i are the same in phenotypic effect, as it makes no difference which allele comes from the mother and which one from the father. Thus, we consider these two genotypes identical. Some alleles may be *dominant* over others alleles, which we call *recessive*. If A_i is dominant over recessive A_j, then the phenotype of the individual is determined by A_i and the presence of A_j remains latent, so that individuals of genotype A_iA_i and A_iA_j are of the same phenotype (with respect to locus A), and only individuals of genotype A_jA_j manifest the recessive trait. It is to be noted that here and in the sequel we consider only so-called *autosomal* chromosomes; *sex-linked* chromosomes that determine the sex of the offspring have some asymmetry and we are not dealing with them here.

Mature individuals produce haploid gametes by a specific and very precisely controlled mechanism called *meiosis*, starting from diploid germ cell precursors. Meiosis involves the reshuffling of alleles between the homologous chromosomes of the germ cell precursor and then the separation of the homologs into haploid gametes. The process itself is very complicated but the result is simple: in effect, meiosis splits the diploid genome of the parental zygote into two haploid (gametal) chromosome sets. Speaking on the locus level, the two members of the allele pair A_iA_j find themselves in two different gamete cells after meiosis. This process is accompanied by other phenomena to which we shall return later.

For locus A with n different alleles the number of possible genotypes is $(n+1)n/2$. We may consider several, say, k loci together. If we assume for the sake of simplicity that n alleles belong to each then, obviously, we may distinguish $((n+1)n/2)^k$ genotypes. This number will increase very rapidly as we increase the number of loci we want to take into consideration. The number of all loci is very large even in the case of the most primitive organisms and this is especially so in higher organisms such as mammals. That is the reason why each human is unique–no two of us have the same overall genotype apart from single zygote twins. The number of individuals in any real population is "infinitesimally small" compared to the number of possible overall genotypes.

Now we shall consider the simplest possible meaningful case, that of a single locus with two alleles, A_1 and A_2 . At a given moment in the population the number of zygotes of genotype $A_1A_1, A_1A_2 = A_2A_1, A_2A_2$ will be denoted by x_{11}, x_{12}, x_{22} , respectively. Thus, the number of the total zygote population is $\overline{x} = x_{11} + x_{12} + x_{22}$. The frequencies of these genotypes are $x = x_{11}/\overline{x}$,

$y = x_{12}/\overline{x}$, $z = x_{22}/\overline{x}$, so that

$$x + y + z = 1, \qquad x, y, z \geq 0 . \tag{4.1.1}$$

The number of gametes is $2\overline{x}$, the number of gametes carrying the allele A_1 or A_2 respectively, is $2x_{11} + x_{12}$ or $2x_{22} + x_{12}$, respectively. Thus, the frequency of A_1-, A_2-gametes respectively, (in the gamete population) is

$$p = \frac{2x_{11} + x_{12}}{2\overline{x}} = x + y/2 \,, 1 - p = \frac{2x_{22} + x_{12}}{2\overline{x}} = z + y/2 \,, \tag{4.1.2}$$

respectively. It is clear from this formula that the zygote frequencies x, y, z determine the gamete frequencies p and $1-p$ uniquely but the latter ones, that is p, in general, does not determine x, y, z uniquely; however, under certain restrictions it does.

The Hardy-Weinberg Law. Assume that: (i) there is random mating in the population, that is, any gamete has an equal chance to unite with any other gamete; (ii) the numbers are sufficiently high so that the frequencies can be identified with the probabilities or, in other words, the probability that a randomly chosen gamete from the pool is of genotype A_1 is p; (iii) the generations are discrete; (iv) the locus, the gene site we are dealing with, is not sex linked; and (v) Darwinian selection does not act in the population, that is, each zygote has an equal chance to survive to adulthood and reproduce. *Under conditions (i)-(v) the genome type frequencies of gametes are constant from generation to generation and from the second generation on the genotype frequencies of zygotes are also constant.* Indeed, if we consider the genotype frequencies established here as those of the parent generation, then at reproduction the chance that an A_1-gamete unites with an A_1-gamete is p^2, that an A_1-gamete unites with an A_2-gamete is $2p(1-p)$ (as the egg may be A_1 and the sperm A_2 or vice versa), and that an A_2-gamete unites with an A_2-gamete is $(1-p)^2$. Thus, the frequencies of genotypes $A_1A_1, A_1A_2 = A_2A_1, A_2A_2$ in the offspring zygote population are

$$x_2 = p^2, \qquad y_2 = 2p(1-p), \qquad z_2 = (1-p)^2 \,, \tag{4.1.3}$$

respectively, $(x_2 + y_2 + z_2 = 1)$, and as a consequence, the frequency of A_1 gametes in the second generation is

$$x_2 + y_2/2 = p^2 + p(1-p) = p \,,$$

as stated. From this second generation on this constant gamete frequency determines the constant zygote genotype frequencies by Eq. (4.1.3). If the A_1 gene frequency $p \in (0,1)$ is varied Eq. (4.1.3) determines a curve on the simplex (4.1.1). This curve is called the *Hardy-Weinberg parabola* (see Fig. 4.1.1) . Only those points of the simplex (4.1.1) may represent the zygote frequencies that lie on this parabola.

The Hardy-Weinberg law is also valid when more than two alleles belong to the given locus. Denoting the alleles by A_i, $(i = 1, 2, ..., n)$ and their respective frequencies by p_i, $(p_1 + p_2 + ... + p_n = 1)$, these remain unchanged, and the frequencies p_i^2 of the "homozygous genotypes" A_iA_i and $2p_ip_j$ the "heterozygous genotypes" A_iA_j also stay constant from the second generation on. This can also been proved easily.

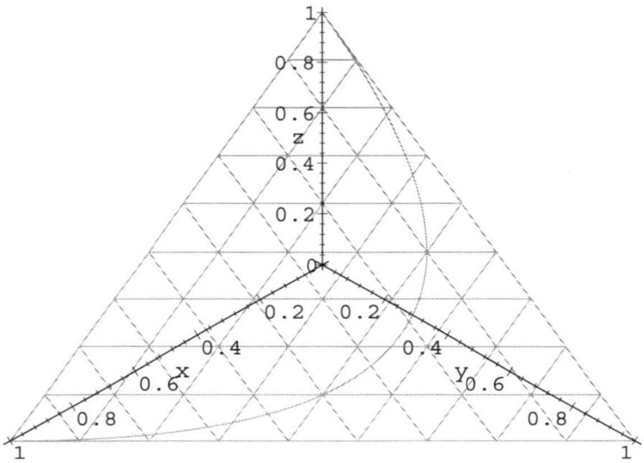

Figure 4.1.1: The Hardy-Weinberg parabola of zygote frequencies (MAPLE).

As an example, note that the frequency of albinos in the human population is ≈ 0.00005 (i.e., there is ≈ 1 albino among 20,000 people). There is a dominant allele A_1 for normal pigmentation and a recessive allele A_2 responsible for albinism. The individual becomes an albino only if he/she develops from a homozygote A_2A_2 and is of normal pigmentation if at the given locus he/she has A_1A_2 or A_1A_1. Denoting the frequencies of the homozygote A_1A_1, the heterozygote A_1A_2 and the albino homozygote by x_2, y_2, z_2, respectively, by Eq. (4.1.3) the frequency of the albino gene A_2 is $1 - p = \sqrt{z_2} = \sqrt{0.00005}$ $= 0.00707$, and that of the dominant gene A_1 is $p = 0.99293$. Hence, the frequency of those who carry the albino gene is

$$y_2 + z_2 = 2p(1-p) + (1-p)^2 = 2 \times 0.99293 \times 0.00707 + 0.00005 = 0.01409,$$

that is, it is well above 1%.

We deal with natural selection in the next Section. However, it must be noted here that selection does not act much on recessive genes because in heterozygotes recessive genes do not affect the manifest phenotype of the individual. Based on the hypotheses of the Hardy-Weinberg law the gene or gamete pool has been considered static in this Section. In the next Section effects of natural selection, recombination (or crossing over), and mutation will be taken into account and we shall also drop the "nonoverlapping generations assumption." As a consequence, we shall deal with n genome types of gametes whose frequencies in the total gamete population at time t will be denoted by $p_i(t)$, $(1, 2, ..., n)$. The phase space in which the dynamics, the evolution of the gene pool, takes place will be the simplex

$$S = \{x = (x_1, x_2, ..., x_n) \in \mathbb{R}^n : x_1 + x_2 + ... + x_n = 1,\ x_i \geq 0,\ i = 1, 2, ..., n\} \tag{4.1.4}$$

of the n-dimensional space because $p_1(t) + p_2(t) + ... + p_n(t) = 1$ always.

4.2 Equations of Selection, Recombination, and Mutation

In this Section we first take into consideration the effect of natural selection upon the frequency distribution of genotypes and prove Fisher's Fundamental theorem of population genetics (see Fisher, 1930), and then we shall build the effects of recombination (crossing over) and mutation into the equations. Finally, we are going to discuss Kimura's Steepest Ascent law (see Crow-Kimura, 1970).

We divide the population of gametes into n genome types and assume that the numbers and frequencies of these genome types are smooth functions of (continuous) time t. The number of genome type i gametes at time t is denoted by $x_i(t)$ and their frequency by $p_i(t)$, $(i = 1, 2, ..., n)$. The number of genotype ij zygotes (zygotes that are formed by the genome type i and the genome type j gamete) is denoted by $x_{ij}(t)$ and their frequency by $p_{ij}(t)$. We denote the total number of zygotes (dropping the argument t) by $\overline{x} = \sum_{i,j=1}^{n} x_{ij}$, so that the total number of gametes is $2\overline{x} = \sum_{i=1}^{n} x_i$. Thus, the gamete frequencies are $p_i = x_i/2\overline{x}$, that is, $x_i = 2\overline{x}p_i$. Applying the Hardy-Weinberg law for one generation gives zygote frequencies of $p_{ij} = x_{ij}/\overline{x} = 2p_ip_j$ if $i \neq j$, and $p_{ii} = x_{ii}/\overline{x} = p_i^2$. Hence,

$$x_{ij} = 2\overline{x}p_ip_j = x_ip_j \ , \ i \neq j \ , \qquad x_{ii} = \overline{x}p_ip_i = p_ix_i/2 \, . \tag{4.2.1}$$

We posit that different zygotes reproduce with different levels of success. During reproduction a zygote of genotype ij produces two gametes, one of genome type i and another of genome type j. These gametes then unite with other gametes to form new zygotes. Because we count gametes we may assume that zygote of type ij gives birth to zygote of type ij. We denote the per capita birth rate of ij zygotes by b_{ij} and their death rate by d_{ij}, and these we consider constant under constant external conditions. Thus, the per capita net reproduction rate of zygotes of type ij is $m_{ij} = b_{ij} - d_{ij}$. We call this the *fitness* of genotype ij. The difference of fitnesses is what counts in the race for survival or dominance within the pool of different genotypes. If i is fixed then the increase of genome type i gametes in unit time is obtained from the reproduction of all the zygotes that contain a gamete of type i . If the zygote contains just one gamete of type i, then its contribution is its fitness. If it is a homozygote of type ii then its contribution is a double fitness, so that the following differential equation is obtained for the number of gametes of type i :

$$\frac{dx_i}{dt} = 2m_{ii}x_{ii} + \sum_{j \neq i} m_{ij}x_{ij} \ , \qquad i = 1, 2, ..., n \, . \tag{4.2.2}$$

Therefore, we want to have a system for the gamete frequencies, in which we substitute from Eq. (4.2.1)

$$\frac{dx_i}{dt} = x_i m_{ii} p_i + x_i \sum_{j \neq i} m_{ij} p_j = x_i \sum_{j=1}^{n} m_{ij} p_j = m_i(p) x_i \ , \ i = 1, 2, ..., n \tag{4.2.3}$$

where the expression

$$m_i(p) := \sum\nolimits_{j=1}^{n} m_{ij} p_j$$

is obviously the weighted average fitness of the zygotes that contain a gamete of type i. Unlike m_{ij} it is not a constant, but depends on the frequencies $p = (p_1, p_2, ..., p_n)$, which vary in time. We may consider m_i the fitness of the gamete of type i. We introduce also the weighted *average fitness* of the population:

$$\overline{m}(p) := \sum_{i,j} m_{ij} p_i p_j = \sum\nolimits_{i=1}^{n} m_i(p) \ p_i \ .$$

Note that we consider zygotes of type ij and ji of the same genotype; this means that in this quadratic form $m_{ij} = m_{ji}$ and for the heterozygotes, for instance, $m_{12} p_1 p_2 + m_{21} p_2 p_1 = 2 m_{12} p_1 p_2 = m_{12} p_{12}$; for the homozygotes it is $m_{ii} p_i^2 = m_{ii} p_{ii}$. To obtain a differential equation for the frequencies we need the derivative of the total number of the population. Adding up the equations (4.2.3) on the left-hand side we obtain the derivative of the sum of the gamete quantities, which is the double of the total zygote population:

$$\sum_i \frac{dx_i}{dt} = 2 \frac{d}{dt} \overline{x} = \sum_i m_i x_i = 2 \overline{x} \sum p_i m_i = 2 \overline{x} \, \overline{m} \, ,$$

that is,

$$\frac{d\overline{x}}{dt} = \overline{m} \, \overline{x} \ . \tag{4.2.4}$$

Now,

$$\begin{aligned} \frac{dp_i}{dt} &= \frac{d}{dt} \left(\frac{x_i}{2\overline{x}} \right) = \frac{1}{2} \frac{\dot{x}_i \overline{x} - x_i \, \dot{\overline{x}}}{\overline{x}^2} = \frac{1}{2} \frac{m_i x_i \overline{x} - x_i \overline{m} \overline{x}}{\overline{x}^2} \\ &= m_i \frac{x_i}{2\overline{x}} - \frac{x_i}{2\overline{x}} \overline{m} = p_i \left(m_i - \overline{m} \right) , \end{aligned}$$

where Eqs. (4.2.3) and (4.2.4) were applied. Finally, substituting the expressions for the average fitnesses we obtain *Fisher's differential equation* or the *Selection equation* for the dynamics of genome type frequencies:

$$\frac{dp_i}{dt} = p_i \left(\sum_j m_{ij} p_j - \sum_{k,j} m_{kj} p_k p_j \right) , \qquad i = 1, 2, ..., n \ . \tag{4.2.5}$$

From this equation one can see that if the fitness of genome type i gametes is less than the average fitness of the population then their frequency in the pool of gametes decreases and if their fitness is above average their frequency grows. As the sum of the frequencies is one, this observation suggests that higher fitnesses will have higher weight and lower ones will have lower weight as time passes and thus one may infer that the average fitness of the population should increase. This is, indeed, the *Fundamental Theorem of Population Genetics: Along the trajectories of the selection equation the average fitness is an increasing function of time.* We shall prove this theorem now. We need the time derivative of the average fitness:

$$\begin{aligned} \dot{\overline{m}} &= d\left(\sum m_{ij}p_ip_j\right)/dt = \sum m_{ij}\left(\dot{p}_ip_j + p_i\dot{p}_j\right) \\ &= \sum m_{ij}\left(p_i\left(m_i - \overline{m}\right)p_j + p_ip_j\left(m_j - \overline{m}\right)\right) \\ &= 2\sum m_{ij}m_ip_ip_j - 2\overline{m}\sum m_{ij}p_ip_j = 2\left(\sum_i m_i^2p_i - \left(\sum_i m_ip_i\right)^2\right), \end{aligned}$$

where in the last step in the first term we performed the summation for j and in the second term we applied this form of $\overline{m}$. One may arrive at the last expression the following way–we may consider fitnesses m_i, $(i = 1, 2, ..., n)$ of the gametes as values of a random variable, which are assumed with probability p_i ; then the *variance* of this probability distribution is

$$\begin{aligned} D^2 &= \sum_i p_i\left(m_i - \sum_j p_jm_j\right)^2 \\ &= \sum_i p_i\left(m_i^2 - 2m_i\sum_j p_jm_j + \left(\sum_j p_jm_j\right)^2\right) \\ &= \sum_i p_im_i^2 - 2\left(\sum_i m_ip_i\right)^2 + \left(\sum_j p_jm_j\right)^2 \\ &= \sum_i p_im_i^2 - \left(\sum_i p_im_i\right)^2 . \end{aligned}$$

The last expression is half of what is found in the expression for $\dot{\overline{m}}$, so that

$$\dot{\overline{m}} = 2D^2 . \qquad (4.2.6)$$

As we can see, the derivative of the average fitness of the population is positive unless the variance is zero, which means that all genotype fitnesses are equal (to the average); if we look at the selection differential equation this happens only if the system is in equilibrium. This proves the Fundamental theorem.

It is seen from Eq. (4.2.6) that the increase of average fitness in unit time is proportional to the variance of the fitnesses of the existing genotypes. This means that the greater the difference between the fitnesses the faster the average fitness grows, and the faster the frequency of low fitness genotypes decreases and that of genotypes with high fitness increases. In the process some genotypes may die out and the system may reach a frequency distribution that corresponds to a *local maximum* of the function $\overline{m}$. For more detailed discussion of the possibilities see Hofbauer-Sigmund (1988). The graph of the function $\overline{m}$ defined over the simplex $\{p = (p_1, p_2, ..., p_n) \in \mathbb{R}^n_+ : p_1 + p_2 + ... + p_n = 1\}$ is called the *fitness landscape*. It plays an important role in modern theories of evolution (see, e.g. Kauffman, 1993).

Besides natural selection there are other phenomena that play an important role in the formation of the genotype pool of subsequent generations. The most important ones are *recombination* (or *crossing over*) and *mutation.*

At meiosis a partial exchange of alleles between the homologous chromosomes may (and usually, does) take place. This process results in the partial reshuffling of genes in the gametes that developed from the same precursor cell, which thus induces genetic variability on the genome-type level. Consider the gametes in question as the union of genes at the loci that belong to a subset S of all loci on one hand, and those at loci of the complement subset $\overline{S}$, and then suppose that the genes of the two gametes that sit on the homologous loci of S change places. (When we say "all the loci" we naturally mean, only those loci that are taken into account in the division of the population into genotypes.) As a consequence, the meiosis of a heterozygote of type ij does not yield an $i-$ gamete and a $j-$ gamete but a $k(i,j)$ and an $l(i,j)$ gamete–the first being of type $k = \widehat{i_S j_{\overline{S}}}$, that is, it has the genes of i at the loci of S and the genes of j at the loci of $\overline{S}$, and the other one is of type $l = j_S i_{\overline{S}}$. See Fig. 4.2.1.

By applying the formulae preceding Eq. (4.2.1), the number of $i-$gametes produced by ij zygotes in unit time is $b_{ij}x_{ij} = b_{ij}\overline{x}p_{ij} = 2\overline{x}b_{ij}p_ip_j$. Denoting the probability of crossing over as described in the preceding text by r^S_{ij}, the resulting loss of $i-$gametes in unit time is $r^S_{ij}2\overline{x}b_{ij}p_ip_j$. However, crossing over in meiosis of a $k(i,j)$ $l(i,j)$ zygote may yield an $i-$gamete; the gain in $i-$ gametes in unit time is $r^S_{kl}2\overline{x}b_{kl}p_kp_l$. We say that the model is *completely symmetric* if the two recombination probabilities are equal, $r^S_{ij} = r^S_{k(i,j)l(i,j)}$ and the birth rates are also equal, $b_{ij} = b_{k(i,j)l(i,j)}$. (This last condition can be justified by considering that the two zygotes have exactly the same genes but on different gametes.) Under this assumption the total loss or gain in $i-$ gametes in unit time is

$$\left.\frac{dx_i}{dt}\right|_{\text{recomb}} = -2\overline{x} \sum_{j \neq i, S} r^S_{ij} b_{ij} (p_i p_j - p_{k(i,j)l(i,j)})$$

where summation is over all j different from i and for all possible choice S of a subset of loci considered. If we sum up these quantities for i then because of the symmetry in i and j the sum is zero, i.e. recombination does not alter the total number of gametes (in unit time). As a consequence, the decrease of the

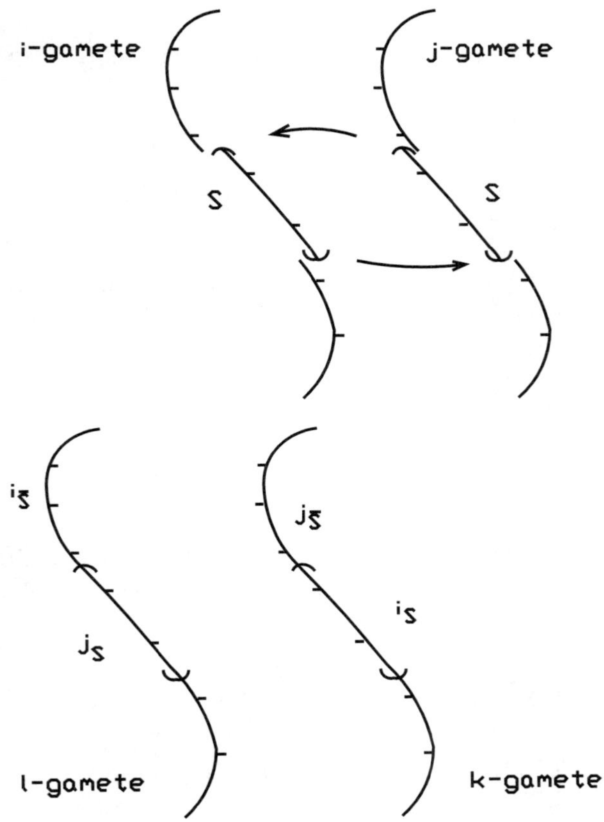

Figure 4.2.1: Crossing over. Recombination of an $i-$ and a $j-$ gamete into a $k-$ and an $l-$ gamete.

frequency of i-gametes in unit time due to recombination is

$$\left.\frac{dp_i}{dt}\right|_{\text{recomb}} = -\sum_{j\neq i,S} r_{ij}^S b_{ij}\left(p_i p_j - p_{k(i,j)l(i,j)}\right). \tag{4.2.7}$$

The genes at different loci of the gamete may be replaced by other genes due to some external effects, such as radioactive radiation, cosmic rays etc. This replacement of genes is called *mutation.* Mutation was considered a random effect until quite recently, when a deeper understanding of genetic structure made it possible to perform artificial mutation in the laboratory and this led to fantastic perspectives and dangers in the development of the science of biology. Mutation has a fundamental role in evolution; without it the abundance of species could not be explained and the effect of Darwinian natural selection would have a very limited playground. Denote the probability that an $i-$ gamete is transformed into a $j-$ gamete in unit time by n_{ij}. The increase and decrease, respectively, of the number x_i of $i-$gametes in unit time due to mutation comes from the gain due to mutation of arbitrary $j \neq i$ gametes into $i-$ gametes and from the loss due to mutation of existing $i-$ gametes into some $j \neq i$ gamete, thus,

$$\left.\frac{dx_i}{dt}\right|_{\text{mutation}} = \sum_{j\neq i} x_j n_{ji} - x_i \sum_{j\neq i} n_{ij} = 2\overline{x}\left(\sum_{j\neq i} p_j n_{ji} - p_i \sum_{j\neq i} n_{ij}\right) .$$

If we sum up these formulae for i then we get zero, that is, the change in the number of gametes in unit time due to mutation is $d2\overline{x}/dt = 0$. Hence, dividing by $2\overline{x}$ and introducing the notation $n_i = \sum_{j\neq i} n_{ij}$ (this is the probability that an $i-$ gamete becomes a gamete of some other type in unit time) for the *frequency* of the gametes we obtain

$$\left.\frac{dp_i}{dt}\right|_{\text{mutation}} = \sum_{j\neq i} p_j n_{ji} - p_i n_i . \tag{4.2.8}$$

If Eqs. (4.2.5), (4.2.7), and (4.2.8) are added together, taking into account the effects of selection, recombination, and mutation, this system of equations is obtained

$$\begin{aligned}\frac{dp_i}{dt} &= p_i\left(\sum_j m_{ij}p_j - \sum_{k,j} m_{kj}p_k p_j\right) - \sum_{j\neq i,S} r_{ij}^S b_{ij}(p_i p_j - p_{k(i,j)l(i,j)}) \\ &\quad + \sum_{j\neq i} p_j n_{ji} - p_i n_i \qquad , \qquad (i = 1, 2, ..., n) .\end{aligned} \tag{4.2.9}$$

We have to observe that very little can be done with this "master equation." Detailed discussion about it can be found in Hofbauer and Sigmund (1988) and Akin (1979). While it would be difficult mathematically to handle it there are important restrictions built into this model. We mention just two of these. First of all, the zygote population has no age structure; that is, the moment a new zygote is born it is able to reproduce. Second, only a very limited number of loci can be taken into consideration. If we want to identify the frequencies of genotypes with the probabilities of their occurrence in the population (with high probability), that is, we want to apply, tacitly, "laws of large numbers," the number of different genotypes n must be low compared to the total population, so that the volume of a typical genotype be large. We count here the genome types of *gametes* (not zygotes). For example, in case of humanity for the sake of simplicity we take the number of humans equal to 5 billions, and this means that there are 10 billion $= 10^{10}$ gametes present in the population. If we consider $n =$ 10 million $(= 10^7)$ different genome types, then the average number of individual gametes in a genome type is 1000. If the number of loci taken into consideration is denoted by k and we assume that 10 alleles belong to each locus, then the number of different genome types is $n = 10^k$, that is, the number of loci that can be taken into consideration is $k = 7$. This is not much compared to the thousands of loci on human chromosomes. Even if we assume that only two alleles belong to each locus then $n = 2^k = 10^7$, from where $k = 7/\log_{10}(2) = 23.253 \approx 23$. On the other hand, we see that the dimension of system (4.2.9) is fairly high.

In addition to Fisher's Fundamental theorem, which says that along evolutionary paths the average fitness of the population is increasing, population genetics has another basic principle, usually called *Kimura's Maximum principle,* but the essence is expressed better if it is called *Kimura's Steepest Ascent principle.* It postulates that the average fitness is not just an increasing function of time but evolution is such that *at every moment the frequency distribution is changing in the direction in which the average fitness is increasing the fastest.* The average fitness $\overline{m}(p) := \sum_{i,j} m_{ij} p_i p_j$ is a scalar function of the frequencies p_i, $(i = 1, 2, ..., n)$, which vary on the simplex S (see Eq. (4.1.4)). The directional derivative of a scalar function is maximal, that is, the function is increasing the fastest in the direction of its gradient vector $\nabla \overline{m} = [\overline{m}'_{p_1}, \overline{m}'_{p_2}, ..., \overline{m}'_{p_n}]$; this is the direction of the steepest ascent. If this vector points out of the simplex S then we have to take its projection on S. Thus, *Kimura's principle says that the trajectories of the system of differential equations, which governs the dynamics of the frequency distribution must go in every point in the direction of the vector $\nabla \overline{m}$ (or its projection onto S), or in other words, the tangent vector of the trajectories must be parallel to this gradient everywhere, that is, the vector field on the right-hand side of the differential system must be parallel to $\nabla \overline{m}$.*

We shall go into detail about the fruitless efforts spent trying to prove the validity of this principle on the basis of the "master equation" (4.2.9) but return to the pure Selection equation (4.2.5). Because $\overline{m}'_{p_k}(p) = 2\sum_j m_{kj} p_j$, the gradient of $\overline{m}$ is

$$\nabla \overline{m}(p) = 2\left[\sum_j m_{1j} p_j, \sum_j m_{2j} p_j, ..., \sum_j m_{nj} p_j\right].$$

The normal vector of S is the vector $u = [1, 1, ..., 1]$. If we take the projection of $\nabla \overline{m}$ to this vector and then subtract the result from $\nabla \overline{m}$ we obtain the projection of $\nabla \overline{m}$ to S. Leaving the details of the calculation to the reader as an exercise we obtain for this projection

$$\begin{aligned}\nabla \overline{m}_S(p) &= \nabla \overline{m}(p) - (\nabla \overline{m}(p) \cdot u / |u|)\, u / |u| \\ &= 2\Bigg[\sum_j m_{1j} p_j - \sum_i m_i/n, \sum_j m_{2j} p_j - \sum_i m_i/n, \ldots, \\ &\qquad \sum_j m_{nj} p_j - \sum_i m_i/n\Bigg].\end{aligned}$$

This vector is *not* parallel to the vector field on the right-hand side of Eq. (4.2.5). One has to discard either Kimura's fairly reasonable principle or system (4.2.5), which was constructed to model pure selection without any artificial condition and has been successful in proving the Fundamental theorem. If one does not want to give up either, one may be led to the conclusion that perhaps the geometry of the phase space of the quantities $x = (x_1, x_2, ..., x_n)$ of different genotypes is not the right one; it is possible that in this phase space the distance of different states is not to be measured the Euclidean way (using the Pythagorean theorem). And, indeed, this has been Shahshahani's solution to the problem. In the interior of the positive orthant $\mathbb{R}^n_+ = \{x \in \mathbb{R}^n : x_i \geq 0, i = 1, 2, ..., n\}$ of the space, *Shahshahani's metric* is introduced (see Appendix 4.2) according to which the square of the distance of the "neighboring points" x and $x + dx = (x_1 + dx_1, x_2 + dx_2, ..., x_n + dx_n)$ is

$$ds^2 = \frac{1}{x_1}dx_1^2 + \frac{1}{x_2}dx_2^2 + ... + \frac{1}{x_n}dx_n^2 \,, \qquad x_i > 0 \,,\; i = 1, 2, ..., n \,. \qquad (4.2.10)$$

The meaning of this metric is that the change dx_i of the quantity x_i is weighted by the (square root of the) reciprocal of its actual value, reflecting the fact that if the actual quantity is small then a change in it is much more important than it would be when the quantity is large. This way a Riemannian metric is introduced in the phase space and computations are made according to the rules of Riemannian geometry (see Appendix 4.1). Riemannian geometry is the generalization of the geometry of 2D curved surfaces to n dimension. However, as it may happen that a 2D surface given in the 3D Euclidean space by some complicated equation turns out to be a plane, that is, a manifold where Euclidean geometry prevails, an $n-$ dimensional Riemannian space where the metric is given by some quadratic form such as Eq. (4.2.10) may turn out to be a Euclidean one. This is the case if by a suitable coordinate transformation the metric, the square of the distance of "neighboring points," can be transformed into a pure (Pythagorean) sum of the squares of the coordinate differences. And, indeed, if the coordinate transformation

$$x_i = \widetilde{x}_i^2/4 \,, \qquad dx_i = \frac{\partial x_i}{\partial \widetilde{x}_i} d\widetilde{x}_i = \frac{1}{2}\widetilde{x}_i d\widetilde{x}_i \,, \qquad i = 1, 2, ..., n \qquad (4.2.11)$$

is performed, then Eq. (4.2.10) is transformed into

$$ds^2 = d\widetilde{x}_1^2 + d\widetilde{x}_2^2 + ... + d\widetilde{x}_n^2 \,,$$

which means that even if the Shahshahani metric (4.2.10) is accepted, the interior of $\mathbb{R}^n_+$ still remains an $n-$dimensional Euclidean space but the coordinates $(x_1, x_2, ..., x_n)$, the quantities of the different genotypes, are no longer Cartesian coordinates. In this space Cartesian coordinates are the ones introduced by Eq. (4.2.11), that is, twice the square roots of the quantities : $\widetilde{x}_i = 2\sqrt{x_i}$, $i = 1, 2, ..., n$.

What we are really interested in is the geometry on the simplex S of the *frequencies* p_i whose equation in the coordinates x_i is $x_1 + x_2 + ... + x_n = 1$, $x_i > 0$, $i = 1, 2, ..., n$. Its equation in the Cartesian coordinates $\widetilde{x}_i$ is

$$\widetilde{x}_1^2 + \widetilde{x}_2^2 + ... + \widetilde{x}_n^2 = 4 \,,\; \widetilde{x}_i > 0, \;\; i = 1, 2, ..., n \,,$$

that is, in the Shahshahani metric it is the part of the $(n-1)$ – dimensional sphere with center in the origin and radius 2 in the positive orthant of the space. Now, if we transform the average fitness into the new coordinate system:

$$\widetilde{\overline{m}}(\widetilde{x}) = \sum_{i,j} m_{ij} x_i(\widetilde{x}) x_j(\widetilde{x}) = \frac{1}{16} \sum_{i,j} m_{ij} \widetilde{x}_i^2 \widetilde{x}_j^2 , \qquad (4.2.12)$$

take the gradient of this function, project it to the tangent space of the sphere S at the point $\widetilde{x}$ of the sphere, and transform it into the original coordinate system x, we obtain exactly the vector field on the right-hand side of the Selection equation (4.2.5). The detailed calculation is given in Appendix 4.2. Thus, *in the Shahshahani metric in every point the trajectories go in the direction of the gradient of* $\overline{m}$ *, and this means that the average fitness is increasing the fastest along the trajectories of the Selection Equation.*

4.3 The Origins of Life and Eigen's Hypercycle

The common feature of all living creatures on Earth is that besides maintaining themselves by processing energy (*metabolism*) they are able to reproduce, offspring that are more or less true copies of themselves. The genetic information that characterizes each individual is coded in large deoxyribonucleic acid (DNA) (and for some viruses in ribonucleic acid (RNA)) molecules. Sections of these molecules are the genes that determine the genotype of the individual. The cytoplasm of each prokaryotic and the nucleus of each eukaryotic cell carry this molecule. The DNA molecules are extremely long double sequences of four kinds of nucleotides (guanine, cytosine, adenine, and thymine) bound in a certain way to two backbone molecules and the two sequences of nucleotides are bound together lightly by a strict "complementary rule" to form a structure called the double helix (Watson, 1981). In human DNA there are several billions of these nucleotides. Which of the four sit at the successive places determines the genetic structure. A battery of enzymes serves the purpose of producing exact RNA copies of the coding regions (genes) on the DNA molecule. These so-called messenger RNA molecules move to the protein-synthetizing apparatus of the cell, which they instruct to produce the corresponding protein coded in their nucleotide sequence–an enzyme or a structural protein of the organism. This is a very short and "reductionist" account of how reproduction and protein synthetization happens in living creatures on Earth. A more detailed description can be found in Rowe (1994). Once we have a not too short DNA or RNA molecule, evolution following Darwinian natural selection and governed by laws and equations like those in the previous Section may begin. We do not want to say that in this respect everything has been settled and the final word has been uttered (more advanced new evolution theories continue to be developed, see e.g., Kauffman, 1993 and Maynard Smith and Szathmáry, 1995) but in principle *some kind of evolution* is possible. However, in the natural sciences one of the most important open problems of our age is how the first DNA or RNA molecules came into being, that is, how life started. Several conflicting theories exist that can more or less be divided into two large groups, that of "protein first" and of "RNA first," where the basic problem is that without sufficiently long RNA molecules it is difficult to imagine how complex protein molecules can be synthetized, and without enzymes long DNA or RNA molecules cannot emerge and survive through several generations. We shall elaborate on this last point and then a theory will be dealt with that tries to solve the problem.

Suppose that a polynucleotide molecule is of length m, i.e. its essence is a sequence of m nucleotides, and during its lifetime on the average it produces s copies of itself among which some may not be exact. For the sake of simplicity we assume that all copies are produced at the end of the average lifetime, so that we have discrete generations of polynucleotides and that every nucleotide in the sequence is copied correctly with the same probability $0 < p < 1$. This means that this polynucleotide molecule will have sp^m correct copies by the time it disintegrates. As this is obviously a *branching process* (see Appendix 1.4), the "genotype" of the polynucleotide may survive only if $sp^m > 1$, or if

$$m < \frac{\ln s}{-\ln p} \approx \frac{\ln s}{1-p}, \tag{4.3.1}$$

(for p near to 1 we may write $\ln p \approx p - 1$). Experiments suggest that RNA replication without any enzymes has approximately an error probability of $1 - p = 0.05$. If we assume that s is between 2 and 3, for example, then the bound in (4.3.1) implies a sequence of length < 20. This is far too short for protein synthesis. With the help of enzymes present in the cells of living organisms the error probability decreases considerably (by a factor of ten thousand or more). Accordingly, the length of RNA of certain bacteria may be a few million units. The threshold of RNA length given by (4.3.1) is a serious trap for any theory on the origins of life. If the RNA are not sufficiently long, enzymes cannot be synthetized and without enzymes RNA cannot reach the length necessary for enzyme synthetization. This antinomy is often called the *error catastrophe* or the *information crisis.*

A way out of this trap was suggested by Eigen in 1971 (see Eigen-Schuster, 1979, and the references therein). Called the *hypercycle*, it is a mathematically well-founded theory of that which is not unimaginable from the point of view of "natural history." Laboratory experiments show that in the prebiotic ocean molecules of sugars, amino acids, nucleotides and the like, that is, the building blocks of life could be formed and were present as a consequence of cosmic radiation and electrical discharges. Because metallic ions have a catalytic influence upon these, a large variety of short sequences could emerge out of these. These nucleotide sequences may synthetize quasiprotein molecules, which in turn may help to catalyze some other short nucleotide sequences. However, molecules formed this way disintegrate rapidly and are not able to form a relatively stable structure over generations. If we look closely at this situation, it becomes clear that the information carrying molecule I_1 synthetizes the quasicatalyst E_1, the latter helps with replication of the information carrying molecule I_2, this one synthetizes the quasicatalyst E_2, and so on. A chain is formed, and a large variety of these unstable structures float and compete with each other for building blocks in the "primordial soup." It is then possible that the last member E_{n-1} of a chain like this catalyzes an information-carrying molecule I_n such that the product of the latter is a quasicatalyst E_n, which helps replication of the first member I_1 of the chain.

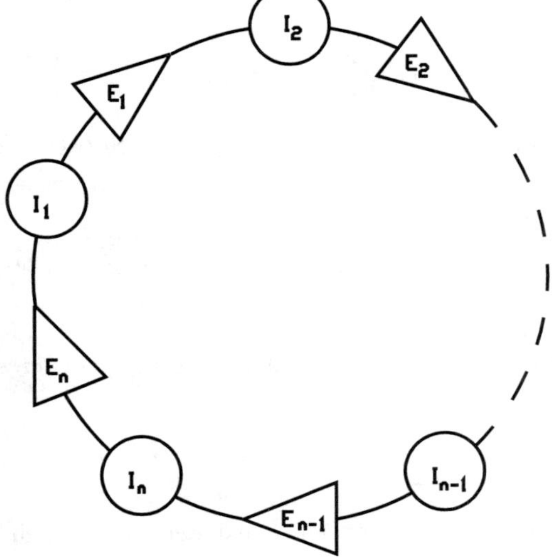

Figure 4.3.1: The hypercycle of nucleotide sequences I_k and quasicatalysts $E_k, (k = 1, 2, ..., n)$.

If this happens a loop is formed (see Fig. 4.3.1), and the members of this loop are in strong selective superiority compared to those outside. The members of the loop may then start a coevolution that leads to life. This is naturally an extremely short account of the essence of the theory. A more detailed exposition can be found in Rowe (1994), where the drawbacks and criticisms of the theory are also put forward. The mathematics of the hypercycle is treated in detail in Hofbauer-Sigmund (1998). We treat here the mathematical theory in a very concise form.

Assuming that the quantity of the kth quasicatalyst is proportional to the quantity of the kth nucleotide sequence, we deal with just n substances where the kth one represents both the kth nucleotide sequence and the quasicatalyst. The basic assumption is that the reproduction of the kth substance is enhanced by the $(k-1)$st, $(k = 2, 3, \ldots, n)$ and that the reproduction of the first substance is enhanced by the nth one. The quantities will be denoted by x_k, $(k = 1, 2, ..., n)$, and in order to simplify the formulae the notation $x_0 = x_n$ will be used. Further, it is assumed that the total quantity in the "soup" is constant, that is, there is a flow that carries away surplus product. The simplest system of differential equations that satisfies these requirements is the following one, written first as expressions for the per capita growth rates

$$\dot{x}_k / x_k = r_k x_{k-1} - g(x) ,$$

where r_k are positive constants and the function g is to be chosen in such a way that $\sum_k x_k$ is constant, that is,

$$\left(\sum_{k=1}^{n} x_k\right)^{\cdot} = \sum_{k=1}^{n} \dot{x}_k = \sum_{k=1}^{n} r_k x_{k-1} x_k - g(x) \sum_{k=1}^{n} x_k = 0 .$$

This means that $g(x) = \sum_{k=1}^{n} r_k x_{k-1} x_k / \sum_{k=1}^{n} x_k$. Substituting this into the system we obtain

$$\dot{x}_k = r_k x_{k-1} x_k - x_k \sum_{i=1}^{n} r_i x_{i-1} x_i \Big/ \sum_{i=1}^{n} x_i . \qquad (4.3.2)$$

Introducing the notations $c = \sum_{i=1}^{n} x_i$, $R_i = c r_i$, dividing the previous equation by c, and turning to the equation for the *frequencies* $p_i = x_i / c$, we get

$$\dot{p}_k = p_k \left(R_k p_{k-1} - \sum_{i=1}^{n} R_i p_{i-1} p_i\right) , \quad k = 1, 2, ..., n . \qquad (4.3.3)$$

This is called the *hypercycle system.* The construction was such that the simplex $S = \{p \in \mathbb{R}_+^n : p_1 + p_2 + ... + p_n = 1\}$ is invariant with respect to this system. Ever since the set up of this system the aim of the research has been to clarify whether it has stable equilibria but if not, then is it able to survive some stable way? Only the most fundamental results are given here.

If system (4.3.3) has an equilibrium $\bar{p} = (\bar{p}_1, \bar{p}_2, ..., \bar{p}_n)$ in the interior of the simplex S then its coordinates must satisfy $\bar{p}_{k-1} = P/R_k$ where $P = \sum_{i=1}^{n} R_i \ \bar{p}_{i-1} \bar{p}_i$. As $1 = \sum p_{k-1} = P \sum 1/R_k$, then $P = 1/\sum 1/R_k$, and $\bar{p}_k = \left(R_{k+1} \sum 1/R_i\right)^{-1}$. The coordinate transformation

$$q_k = \frac{R_{k+1} p_k}{\sum_i R_{i+1} p_i} , \qquad p_k = \frac{q_k / R_{k+1}}{\sum_i q_i / R_{i+1}} , \qquad (k = 1, 2, ..., n)$$

carries the simplex S into itself, as can be seen just by adding up the q_k coordinates, and it takes the point $\bar{p}$ into the point $\bar{q} = (1/n, 1/n, ..., 1/n)$. By a somewhat lengthy but simple calculation system (4.3.3) is transformed into

$$\dot{q}_k = q_k \left(q_{k-1} - \sum_{i=1}^{n} q_{i-1} q_i\right) \mu(q) , \qquad (k = 1, 2, ..., n) ,$$

where $\mu(q) = \left(\sum q_i / R_{i+1}\right)^{-1}$. The trajectories and the direction of the motion along them does not change if we drop the positive scalar factor μ on the right-hand side of the last system. This way we arrive at the system

$$\dot{q}_k = q_k \left(q_{k-1} - \sum_{i=1}^{n} q_{i-1} q_i\right) , \qquad (k = 1, 2, ..., n) , \qquad (4.3.4)$$

which is equivalent to (4.3.3) and easier to handle in our stability investigation. As we have seen, the center $\bar{q} = (1/n, 1/n, ..., 1/n)$ of the simplex S is the only equilibrium of this system in the interior of the simplex.

We study now the stability of this equilibrium point. For a start, the Lyapunov function $V(q) = \ln(q_1 q_2 ... q_n)$ will be used. This function is clearly negative in the interior of S and tends to $-\infty$ as q tends to the boundary of the simplex. It must assume its maximum in the interior of S. We obtain the maximum point by applying the method of Lagrange multipliers to the conditional extremum problem

$$\max_{q \in S} V(q) = ?.$$

We have to find the zeros of the system

$$\begin{aligned} \frac{\partial}{\partial q_i}\left(\ln(q_1 q_2 \ldots q_n) - \lambda\left(\sum\nolimits_j q_j - 1\right)\right) &= 1/q_i - \lambda = 0\,, \qquad (i = 1, 2, ..., n), \\ \sum\nolimits_j q_j - 1 &= 0\,. \end{aligned}$$

The only zero of this system is obviously $\bar{q}$. Thus, the function V assumes its maximum at the equilibrium point, with its maximal value $V(\bar{q}) = n \ln(1/n)$ $= -n \ln n$, so that the function $W(q) = V(\bar{q}) - V(q) = -n \ln n - \ln(q_1 q_2 \ldots q_n)$ is positive definite with respect to point $\bar{q}$; it is positive everywhere on S except at the equilibrium where it is zero. We determine its derivative with respect to system (4.3.4).

$$\begin{aligned} \dot{W}_{(4.3.4)}(q) &= -\sum\nolimits_{i=1}^{n} \dot{q}_i/q_i = -\sum\nolimits_{i=1}^{n}\left(q_{i-1} - \sum\nolimits_{j=1}^{n} q_{j-1}q_j\right) \\ &= -1 + n\sum\nolimits_{j=1}^{n} q_{j-1}q_j = -\left(\sum\nolimits_{i=1}^{n} q_i\right)^2 + n\sum\nolimits_{j=1}^{n} q_{j-1}q_j \end{aligned}$$

where in the last step a little trick was applied to write 1 in a suitable form. We study the cases $n = 2, 3, 4$ separately.

If $n = 2$ a simple calculation yields $\dot{W}_{(4.3.4)_2}(q) = -(q_1 - q_2)^2 \leq 0$, and it is zero only if $q_1 = q_2 = 1/2$, that is, at the equilibrium $\bar{q}$. As a consequence, by Lyapunov's theorem (see Appendix 2.2) $\bar{q}$ is globally asymptotically stable with respect to the interior of the simplex.

In case $n = 3$ a simple calculation yields

$$\dot{W}_{(4.3.4)_3}(q) = -\frac{1}{2}\left((q_1 - q_2)^2 + (q_2 - q_3)^2 + (q_3 - q_1)^2\right) \leq 0\,,$$

and it is zero only if $q_1 = q_2 = q_3 = 1/3$, that is, at the equilibrium $\bar{q}$. As a consequence again, $\bar{q}$ is globally asymptotically stable with respect to the interior of the simplex.

For $n = 4$ a somewhat more complicated calculation yields

$$\dot{W}_{(4.3.4)_4}(q) = -\frac{1}{2}(q_1 + q_3 - q_2 - q_4)^2 \leq 0\,,$$

which is zero if $q \in \Delta = \{q \in S : q_1 - q_2 + q_3 - q_4 = 0)$. We are to apply now the Barbashin-Krasovskiy theorem (see Appendix 2.2) showing that the plane Δ does not contain positive semitrajectories except the equilibrium point $\bar{q}$. The plane contains a trajectory only if its normal vector is orthogonal to the tangent of the trajectory, that is, to the right-hand side of system $(4.3.4)_4$. The normal vector of the plane is the vector $[1, -1, 1, -1]$, and the condition of orthogonality is

$$\begin{aligned} & 1 \cdot q_1 \left(q_4 - \sum_{i=1}^{4} q_{i-1} q_i \right) - 1 \cdot q_2 \left(q_1 - \sum_{i=1}^{4} q_{i-1} q_i \right) \\ & +1 \cdot q_3 \left(q_2 - \sum_{i=1}^{4} q_{i-1} q_i \right) - 1 \cdot q_4 \left(q_3 - \sum_{i=1}^{4} q_{i-1} q_i \right) \\ = \ & q_4 q_1 - q_1 q_2 + q_2 q_3 - q_3 q_4 + (q_1 - q_2 + q_3 - q_4) \sum_{i=1}^{4} q_{i-1} q_i \\ = \ & (q_1 - q_3)(q_4 - q_2) = 0 \, . \end{aligned}$$

Thus, a trajectory is contained in the plane Δ only if along the trajectory either $q_1 = q_3 = \frac{1}{4}$ or $q_2 = q_4 = \frac{1}{4}$ (since $q_1 + q_3 = q_2 + q_4$ and the sum of the four is 1). However, if $q_1 = q_3 = \frac{1}{4}$ along a trajectory then $\dot{q}_1 = \dot{q}_3 = 0$, and from the first and the third equation of system $(4.3.4)_4$ we have $q_4 = \sum_{i=1}^{4} q_{i-1} q_i = q_2 = \frac{1}{4}$, by also taking into account that their sum must be $\frac{1}{2}$. A similar argument for the second possibility implies that the only trajectory contained in the plane Δ is the equilibrium point $\bar{q} = \left(\frac{1}{4}, \frac{1}{4}, \frac{1}{4}, \frac{1}{4}\right)$. Hence, with the quoted theorem the interior equilibrium is globally asymptotically stable in this case.

We have seen that a hypercycle of two, three or four selfreplicating macromolecules arrives at a globally stable equilibrium in the interior of the simplex of frequencies, and this means that its members survive at a constant level in the long run. If the chain is longer the situation is not so simple. One may prove relatively easily by linear stability analysis that the interior equilibrium of a hypercycle of a dimension ≥ 5 is *unstable*. Nevertheless, Hofbauer et al. (1991) have proved that in such higher dimensional hypercycles the solutions tend to orbitally asymptotically stable *periodic solutions*, that is, such hypercycles also survive not on a constant level but with oscillating frequencies of the members. There is mathematical evidence of the expectation that once a hypercycle occurred in the "primordial soup" the replicating molecules that had not become a member died out and also that if several disjoint hypercycles occurred then competition eliminated all but the fittest.

The evolution of the hypercycles is envisaged the following way. Suppose that there is an "old" hypercycle whose participants have quantities $x_1(t), x_2(t), ..., x_n(t)$ as before, and a "new" hypercycle emerges that has a common element, say, that of index 1 with the old one. The quantities of the participants of the new hypercycle will be denoted by $x_1(t), x_{n+2}(t), ..., x_{n+m}(t)$. It is assumed again that the total quantity of all the participants remains constant, that is,

$$\sum_{i=1, i \neq n+1}^{n+m} x_i(t) \equiv c \, .$$

Then in complete analogy to system (4.3.2) we may write the system governing the dynamics of the two hypercycles in the form

$$\begin{aligned}
\dot{x}_1 &= r_1 x_1 x_n + \tilde{r}_1 x_1 x_{n+m} - x_1 g(x) \\
\dot{x}_i &= r_i x_i x_{i-1} - x_i g(x)\,, \qquad (i = 2, 3, ..., n) \\
\dot{x}_{n+2} &= r_{n+2} x_{n+2} x_1 - x_{n+2} g(x) \\
\dot{x}_{n+j} &= r_{n+j} x_{n+j} x_{n+j-1} - x_{n+j} g(x)\,, \quad (j = 3, 4, ..., m)\,, \qquad (4.3.5)
\end{aligned}$$

where $x = (x_1, x_2, ..., x_n, x_{n+2}, ..., x_{n+m})$, $\tilde{r}_1$ is the growth rate of x_1 in the new hypercycle and

$$\begin{aligned}
g(x) &= \frac{1}{c}\Big((r_1 x_n + \tilde{r}_1 x_{n+m}) x_1 + \sum\nolimits_{i=2}^{n} r_i x_i x_{i-1} \\
&\quad + r_{n+2} x_{n+2} x_1 + \sum\nolimits_{j=3}^{m} r_{n+j} x_{n+j} x_{n+j-1} \Big)\,.
\end{aligned}$$

Multiply the second equation by x_{n+2} and the $(n+2)$nd by x_2 (these are the molecules in the old and in the new hypercycles that are catalyzed by x_1) and subtract the two equations from each other,

$$\dot{x}_2 x_{n+2} - x_2 \dot{x}_{n+2} = (r_2 - r_{n+2}) x_2 x_{n+2} x_1\,.$$

Dividing the last equation by $(x_{n+2})^2$ we obtain

$$\left(\frac{x_2}{x_{n+2}}\right)^{\cdot} = (r_2 - r_{n+2}) x_1 \frac{x_2}{x_{n+2}}\,.$$

Dividing the last equation by x_2/x_{n+2} and integrating we get

$$\frac{x_2(t)}{x_{n+2}(t)} \frac{x_{n+2}(0)}{x_2(0)} = \exp\left((r_2 - r_{n+2}) \int_0^t x_1(\tau) d\tau \right)\,.$$

Suppose that $r_2 < r_{n+2}$. In system (4.3.5) the right-hand side of the equation for x_1 is greater than the right-hand side of the corresponding equation in system (4.3.2) but if the dynamics of x_1 is governed by (4.3.2) then as t tends to infinity $x_1(t)$ tends either to a positive constant or to a positive periodic function depending on how large n is. In both cases $x_1(t)$ stays above a positive constant and so does it, *a fortiori*, if its dynamics is governed by (4.3.5). This implies that in the last equation the right-hand side tends to zero as t tends to infinity, that is, $x_2(t)$ tends to zero. But if $x_2(t)$ tends to zero then the right-hand side of the equation for x_3 becomes negative sooner or later, so that $x_3(t)$ also tends to zero, and so on. Thus, we see that if a new hypercycle is formed with x_1 taking part and in this new one it plays the role of the catalyzer more efficiently then the old hypercycle will die out and the new one will take over. This simple argument shows that hypercycles are apt to a certain kind of evolution. This abstract, prebiotic evolution of hypercycles is different from the Darwinian evolution of living creatures that led to an extreme diversity of different species. There is no branching in this imaginary evolution–there is a single line of fitter and fitter macromolecules that leads to the well-defined single device of replicating DNA and RNA that characterizes all living creatures, that is, life on our planet.

4.4 The Immune System

Most living creatures, human beings in particular, are equipped with a defense system that fights foreign materials (like proteins) called *antigenes* that invade the body. This is called the *immune system.* Its formation and mechanism probably presents, the most complex problem faced by contemporary medicine. In the past 30 years and especially since the acquired immune deficiency syndrome (AIDS) became a worldwide scourge caused by a virus that attacks and destroys the immune system, great effort has gone into studying it with several important points still awating clarification. It is impossible to explore the details of the mechanism and to explain the problems in this limited forum. We give here a very simplified description of how the system works and then an extremely simple mathematical model will be treated. For those who want to go to the heart of the matter, Rowe (1994) has written a concise, understandable, and fairly comprehensive account.

The cells that play a role in the immune response of the organism to an attack by foreign cells or molecules develop from *plasma cells* found in bone marrow. There are two important types of *lymphocytes*, cells into which plasma cells develop–*B cells* that mature in the bone marrow and *T cells* that move to the thymus gland and mature there. These B cells synthetize *antibody* molecules that are able to bind to antigenes and destroy them. Three kinds of T cells are distinguished–killer, helper and suppressor T cells. The first type is capable of destroying cells of the host organism that have been infected by the antigene, the second type helps proliferate B cells when an antigene gets into the organism, the third type suppresses immune reaction. When a foreign cell or protein molecule enters the organism for the first time, it triggers multiplication of lymphocytes, especially those that are able to synthetize antibodies that the given antigene. The human immune system is able to produce $\approx 10^8 - 10^9$ different antibodies. Although this is a very large number, it is not yet clear how the system responds to the virtually infinite number of different antigenes that may enter the organism. Neither do we know yet exactly how the system recognizes that it has already encountered a certain antigene because if the same antigene enters the organism after some time has passed since the first encounter, the response is different, that is, the organism may have developed "immunity." Another problem involves the immune system's ability to distinguish "self" cells from foreign ones and not attack the former. Any description of the "immune network" is also in *statu nascendi.* There are different and sometimes conflicting theories explaining these phenomena but we shall not go into this here. DeLisi (1983) and Perelson (1989) treat some of these problems. In the sequel a schematic model from Marchuk (1983) will be presented. This model is perhaps one of the simplest but has the advantage that it takes into consideration the *delay* inherent in the problem of immune reaction and yet still yields itself to relatively easy mathematical study. More sophisticated models, such as Bell's model can be found in the literature quoted and also in Marchuk's work (1983, 1997).

The virus (antigene), the antibody and the lymphocyte (plasma cell) concentrations at time t will be denoted by $V(t), F(t), C(t)$, respectively. Besides these the model has been developed to describe an attack on a well-defined organ (e.g. that of the hepatitis virus on the liver) and is able to measure the damage to the organ in question at time t, denoting it by $m(t)$. Naturally, V, F, C must be positive and it is supposed that $0 \leq m(t) \leq 1$, the values 0 and 1 meaning a healthy organ and complete damage, respectively. Denoting the reproductive rate of the virus by $\beta > 0$ and the "predation rate" of antibodies by $\gamma > 0$, we obtain the equation governing the dynamics of the antigene in the organism

$$\frac{dV}{dt} = (\beta - \gamma F)V .$$

The normal level of plasma cells in a healthy organism will be denoted by C^*; if the level is above this value it drops and if it is below this it increases. The presence of antigenes and antibodies bound to them stimulates the multiplication of immunocompetent B cells, with the growth rate proportional to FV. The production rate seems to be dependent on the condition of the organ attacked–if it has been severely damaged the production rate falls. Based on this hypothesis the rate of proportionality is assumed in the form $\alpha\xi(m)$ where $\alpha > 0$ and ξ is a positive nonincreasing function $\xi(m) \equiv 1$ for $0 \leq m \leq m^*$ and it decreases to zero in the interval $m^* \leq m \leq 1$, expressing the fact that a slightly damaged organ does not influence the immune system. This way we arrive at the second equation

$$\frac{dC(t)}{dt} = \alpha\xi(m(t))F(t-\tau)V(t-\tau) - \mu_c(C(t) - C^*) ,$$

where $\mu_c > 0$, and $\tau > 0$ indicates the delay, that is, the time needed for formation of immunocompetent B cells. Proceeding, we denote by $\rho > 0, \eta > 0$, and $\mu_f > 0$ the production rate of antibodies by B cells, the number of antibodies needed for the neutralization of an antigene, and the death rate of antibodies, respectively. The dynamics of antibody production is then governed by the equation

$$\frac{dF}{dt} = \rho C - \eta\gamma FV - \mu_f F .$$

Finally, assuming that the damage of the target organ is proportional to the quantity of antigene with rate $\sigma > 0$ and that the recuperation rate of the organ is $\mu_m > 0$ we write

$$\frac{dm}{dt} = \begin{cases} \sigma V - \mu_m m, & 0 \leq m \leq 1 \\ 0, & m > 1 \end{cases} .$$

The last equation expresses the fact that if the target organ has been completely destroyed, the damage process stops. In summation, we have obtained a 4D system of differential equations in which we have to write out the argument t

everywhere because it is a delay system:

$$\begin{aligned}
\frac{dV(t)}{dt} &= (\beta - \gamma F(t))V(t) \\
\frac{dC(t)}{dt} &= \alpha\xi(m(t))F(t-\tau)V(t-\tau) - \mu_c(C(t) - C^*) \\
\frac{dF(t)}{dt} &= \rho C(t) - \eta\gamma F(t)V(t) - \mu_f F(t) \\
\frac{dm(t)}{dt} &= \begin{cases} \sigma V(t) - \mu_m m(t), & 0 \le m \le 1 \\ 0, & m > 1 \end{cases} .
\end{aligned} \tag{4.4.1}$$

We are interested in the solutions of this system for $t \geq 0$ because the antigene attack is supposed to occur at time 0. For a delay system one has to give *initial functions on the interval* $(-\tau, 0)$ as initial values (see Appendix 2.1). In our case this is easy,

$$V(t) \equiv 0, \qquad t < 0, \qquad V(0) = V_0 > 0, \tag{4.4.2}$$

and, as delay is present only in the second equation in the first term, which is zero for $t < 0$, it is sufficient to prescribe the values of the rest of the functions at $t = 0$,

$$C(0) = C_0 > 0, \qquad F(0) = F_0 > 0, \qquad 0 \le m(0) = m_0 < 1 . \tag{4.4.3}$$

Note that by this we assume that there are immunocompetent cells and suitable antibodies present at the beginning. If this is not the case the organism either may be defenseless to the given antigene or by "hypermutation" it may develop a response during the attack (see Rowe, 1994) but capturing this phenomenon requires a much more complex model.

We study now the behavior of the solutions of model (4.4.1) under initial conditions (4.4.2)-(4.4.3). The positivity and existence of the solutions is to be shown first and then the stability of the equilibria will be studied.

Solutions belonging to nonnegative initial conditions remain nonnegative for all $t \geq 0$ in their domain of existence. We establish this property the following way From the first equation of the system we have for $t \geq 0$

$$V(t) = V_0 \exp\left(\int_0^t (\beta - \gamma F(s))ds\right) > 0.$$

For $t \in [0, \tau)$ we have by (4.4.2) $V(t-\tau) \equiv 0$ and therefore in this interval the second equation of the system has the form

$$\frac{dC}{dt} = -\mu_c(C - C^*) .$$

If for some $t_0 \in [0, \tau)$ the function C becomes zero, $C(t_0) = 0$, then at this point $\dot{C}\ (t_0) = \mu_c C^* > 0$, so that C cannot assume negative values. The nonnegativity of F and m on interval $[0,\tau)$ is established similarly. An analogous argument works on the interval $[\tau, 2\tau)$ provided that the solutions are defined on this interval too. Similarly, we may establish the nonnegativity of the solutions on $[n\tau, (n+1)\tau)$ for all $n = 0, 1, 2, 3, \dots$.

Every solution belonging to nonnegative initial values is defined on $[0,\infty)$. We show first that the solutions are defined on $[0,\tau)$. On this interval the following obvious estimates hold for the right-hand sides of system (4.4.1)

$$\dot{V} \le \beta V \quad , \quad \dot{C} = -\mu_c(C - C^*) \quad , \quad \dot{F} \le \rho C - \mu_f F \quad , \quad \dot{m} = \sigma V - \mu_m m \, .$$

This means that the *linear system*

$$\dot{x}_1 = \beta x_1 \quad , \quad \dot{x}_2 = -\mu_c(x_2 - C^*) \, , \quad \dot{x}_3 = \rho x_2 - \mu_f x_3 \, , \quad \dot{x}_4 = \sigma x_1 - \mu_m x_4$$

is a "majorant system" to (4.4.1). This implies that any solution of system (4.4.1) that belongs to the same initial conditions as a corresponding solution of the linear system stays below the latter. As the solutions of the linear system are defined for all $t \ge 0$, and the solutions of Eq. (4.4.1) belonging to nonnegative initial values are nonnegative, the latter are defined on $[0,\tau)$ at least. On the interval $[\tau, 2\tau)$ the following inhomogeneous linear system is majorant to (4.4.1),

$$\dot{x}_1 = \beta x_1, \quad \dot{x}_2 = -\mu_c(x_2 - C^*) + b(t), \quad \dot{x}_3 = \rho x_2 - \mu_f x_3, \quad \dot{x}_4 = \sigma x_1 - \mu_m x_4 \, ,$$

where $b(t) = \alpha F(t-\tau)V(t-\tau)$, which is a known function for $t \in [\tau, 2\tau)$ (we have used the fact that $0 \le \xi \le 1$). Using an argument similar to the preceding one, we may establish the existence of the solution of (4.4.1) on $[\tau, 2\tau)$, and so on.

We turn now to the study of the equilibria. To simplify the calculations we suppose that the value of m falls into the interval $[0, m^*)$, so that in the second equation $\xi(m)$ can be replaced by 1. By making the right-hand sides of system (4.4.1) equal to zero an easy calculation yields the two equilibria of the system

$$E_1 = (V_1, C_1, F_1, m_1) = (0, C^*, \rho C^*/\mu_f, 0)$$

and

$$\begin{aligned} E_2 &= (V_2, C_2, F_2, m_2) \\ &= \left(\frac{\mu_c(\mu_f\beta - \rho\gamma C^*)}{\beta(\alpha\rho - \mu_c\gamma\eta)}, \frac{\mu_f\alpha\beta - \mu_c\gamma^2\eta C^*}{\gamma(\alpha\rho - \mu_c\gamma\eta)}, \frac{\beta}{\gamma}, \frac{\sigma}{\mu_m}\frac{\mu_c(\mu_f\beta - \rho\gamma C^*)}{\beta(\alpha\rho - \mu_c\gamma\eta)} \right), \end{aligned}$$

where the second one is in the positive orthant of the space only if

$$\text{sign}\,(\mu_f\beta - \rho\gamma C^*) = \text{sign}\,(\alpha\rho - \mu_c\gamma\eta) = \text{sign}\,(\mu_f\alpha\beta - \mu_c\gamma^2\eta C^*) \ne 0 \, . \quad (4.4.4)$$

To determine the stability of the equilibria we linearize the system at E_i, $i = 1, 2$. The coefficient matrix of the linearized system is the Jacobian matrix of the right-hand side of (4.4.1) but one has to take into account at which term the delays occurred. The linearized system is

$$\begin{aligned}
\dot{y}_1(t) &= (\beta - \gamma F_i) y_1(t) - \gamma V_i y_3 \\
\dot{y}_2(t) &= \alpha F_i y_1(t-\tau) - \mu_c y_2(t) + \alpha V_i y_3(t-\tau) \\
\dot{y}_3(t) &= -\eta\gamma F_i y_1(t) + \rho y_2(t) - (\eta\gamma V_i + \mu_f) y_3(t) \\
\dot{y}_4(t) &= \sigma y_1(t) - \mu_m y_4(t), \qquad (i = 1, 2) . \qquad (4.4.5)
\end{aligned}$$

We obtain the characteristic equation of this system by substituting $y(t) = e^{\lambda t} s$ where s is a constant column vector, the *eigenvector* corresponding to the *eigenvalue* λ. The condition of the existence of a nonzero solution of this form is the *characteristic equation* of the linear delay system of equations (4.4.5):

$$\det \begin{bmatrix} \beta - \gamma F_i - \lambda & 0 & -\gamma V_i & 0 \\ \alpha F_i e^{-\lambda\tau} & -\mu_c - \lambda & \alpha V_i e^{-\lambda\tau} & 0 \\ -\eta\gamma F_i & \rho & -(\eta\gamma V_i + \mu_f + \lambda) & 0 \\ \sigma & 0 & 0 & -\mu_m - \lambda \end{bmatrix} = 0 \quad .$$

The roots of this equation are the eigenvalues. Note that this is no longer necessarily a polynomial equation (see Appendix 2.2). However, in substituting the coordinates of the equilibrium E_1 we are in luck because the characteristic equation turns out to be

$$(\mu_m + \lambda)(\mu_f + \lambda)(\mu_c + \lambda)(\lambda - (\beta - \gamma\rho C^* / \mu_f)) = 0 \, .$$

The equilibrium E_1 is aymptotically stable if all the eigenvalues have negative real parts (see Appendix 2.2). Because the first three roots of the last equation $-\mu_m, -\mu_f, -\mu_c$ are negative we arrived at the following theorem.

THEOREM 4.4.1. *If $\mu_f \beta < \gamma\rho C^*$ then the equilibrium E_1 is asymptotically stable.*

This equilibrium clearly represents the disease-free state, that is, no viruses are present in the organism ($V_1 = 0$) and the organ in question is not damaged ($m_1 = 0$). The condition of the theorem is intuitively reasonable; it says that if the multiplication rate of the antigene and the disintegration rate of the antibody are small, both the normal plasma cell level and the effectiveness of the antibody to bind to the antigene are high, and the antibody's production rate is large, then the disease-free state is stable. However, the stability criterion based on linearization and eigenvalues ensures only *local* stability but does not tell us how large perturbations from equilibrium are tolerated by the system so that it is able to return to the equilibrium state. In the case discussed here, we are able to give a threshold value of the quantity of attacking antigenes under which the immune system is definitely able to drive the intruding antigene to extinction. We shall suppose that at the beginning of the antigene attack the immunocompetent plasma cell, the antibody level, and the state of the organ in question have a disease-free equilibrium value and only the initial antigene quantity differs from the equilibrial value zero.

THEOREM 4.4.2. *If the condition of Theorem 4.4.1 holds, the initial values are* $C_0 = C_1 = C^*$, $F_0 = F_1 = \rho C^*/\mu_f$, $m_0 = m_1 = 0$, *and*

$$0 < V_0 < \frac{\gamma\rho C^* - \mu_f \beta}{\beta\eta\gamma}, \tag{4.4.6}$$

then the function $V(t)$ *is monotone decreasing on* $(0,\infty)$ *and tends to zero as* t *tends to infinity.*

PROOF. Because of the conditions $\beta - \gamma F_0 = \beta - \gamma\rho C^*/\mu_f < 0$, hence, by the first equation of system (4.4.1) $\dot{V}(0) < 0$. Therefore, also $\dot{V}(t) < 0$ in a right-hand side neighborhood of 0. Suppose that V is not decreasing on the whole interval $(0,\infty)$. Then there is a first $t_0 > 0$ such that $\dot{V}(t) < 0$ for $t \in (0, t_0)$, $\dot{V}(t_0) = 0$, and $\dot{V}(t_1) \geq 0$ for some $t_1 > t_0$ arbitrary near to t_0. However, these inequalities imply that $\beta - \gamma F(t_0) = 0$ (as $V(t_0) > 0$), that is, $F(t_0) = \beta/\gamma$ and $\beta - \gamma F(t_1) \geq 0$, that is, $F(t_1) \leq \beta/\gamma$. This implies that $\dot{F}(t_0) \leq 0$ must hold. We deduce an inequality that contradicts the last one. From the third equation of system (4.4.1) we have

$$\dot{F}(t_0) = \rho C(t_0) - \eta\gamma V(t_0)\beta/\gamma - \mu_f\beta/\gamma\,.$$

However, because the derivative of V is negative on $(0, t_0)$, therefore $V(t_0) < V_0$, and by the second equation of system (4.4.1)

$$\frac{d(C(t) - C^*)}{dt} = \frac{dC(t)}{dt} \geq -\mu_c(C(t) - C^*)\,,$$

so that $C(t) - C^* \geq (C_0 - C^*)\exp(-\mu_c t) = 0$, that is, $C(t_0) \geq C^*$. Applying these inequalities and (4.4.6) we obtain the following estimate for the derivative of F,

$$\dot{F}(t_0) > \rho C^* - \eta\beta V_0 - \mu_f\beta/\gamma > \rho C^* - \eta\beta\frac{\gamma\rho C^* - \mu_f\beta}{\beta\eta\gamma} - \mu_f\beta/\gamma = 0\,,$$

and this yields the contradiction. Thus, we have proved that V is decreasing. Now we prove that it tends to zero. As we now have $V(t) < V_0$ for all $t > 0$ we may estimate again the derivative of F, this time at arbitrary $t > 0$ from the third equation of (4.4.1)

$$\dot{F}(t) > \rho C^* - (\eta\gamma V_0 + \mu_f)F(t)\,.$$

This implies that $F(t)$ is greater than the solution of the differential equation

$$\dot{y}(t) = \rho C^* - (\eta\gamma V_0 + \mu_f)y(t)$$

that belongs to the initial condition $y(0) = F_1 = \rho C^*/\mu_f$, that is, for $t > 0$

$$F(t) > \frac{\rho C^*}{\eta\gamma V_0 + \mu_f} + \left(\rho C^*/\mu_f - \frac{\rho C^*}{\eta\gamma V_0 + \mu_f}\right) e^{-(\eta\gamma V_0 + \mu_f)t} > \frac{\rho C^*}{\eta\gamma V_0 + \mu_f} .$$

By estimating the first equation of (4.4.1) applying the last inequality we obtain

$$\dot{V}(t) < \left(\beta - \gamma\frac{\rho C^*}{\eta\gamma V_0 + \mu_f}\right) V(t) ,$$

and this means that for $t > 0$

$$0 < V(t) < V_0 \exp\left(\beta - \gamma\frac{\rho C^*}{\eta\gamma V_0 + \mu_f}\right) t$$

but by (4.4.6) the expression in brackets is negative, which implies that V tends to zero, indeed, as t tends to infinity.□

The expression on the right-hand side of inequality (4.4.6) is called the *immunological barrier*. The immune system is definitely able to resist an attack if the intruding antigenes are fewer than the barrier. Continuing the explanation given following Theorem 4.4.1, if we look at the denominator of the immunological barrier we see that the fewer antibodies needed for neutralization of an antigene, the higher is the barrier. While the rest of the parameters seem to be difficult to influence it is clear that the barrier can be raised by increasing the normal immunocompetent plasma cell level C^* and the rate of antibody production ρ.

A much more difficult stability analysis is needed for equilibrium E_2, which represents a chronic disease caused by an antigene. We can not go into this here but Marchuk (1983) establishes a sufficient condition for the asymptotic stability of E_2 that uses the reversed inequality of Theorem 4.4.1. This means that there exists a sufficient condition for the stability of E_2 if the sufficient condition of stability of E_1 does not hold. It is an open question whether both E_1 and E_2 could be stable at the same time. In this case the outcome of an attack by an antigene would depend on the initial conditions.

Chapter 5

MORPHOGENESIS AND PATTERN FORMATION

In this chapter we look at some of the attempts to construct mathematical models that could describe mammalian cellular differentiation at the embryonic stage that leads to the formation of different organs. In the first Section, Thom's catastrophe theory, probably the first general mathematical theory that attempted to capture this phenomenon in its entire complexity, will be concisely handled. In the second Section the less general, but notwithstanding, difficult problem of pattern formation in animal hides and shells will be treated.

5.1 Modeling the Embryo–Thom's Catastrophe Theory

Thom's *Stabilité structurelle et morphogénèse* was published in 1972 and resulted in a kind of revolution in applied mathematics. Although dynamical systems had been applied earlier in biology–for instance, by Euler, the Bernoullis, Volterra, Fisher, and others–this book made a breakthrough in applying mathematics in the "soft sciences" including biology, psychology, the humanities etc. To be sure, the revolution was followed by a restoration, as was often the case in other respects, and some applications were severely criticized; nevertheless, since then "the world has been different." In this Section an informal presentation of the elementary catastrophe theory will be given along with those attempts made to apply it to embryology. It is to be observed that this application was Thom's (1972) main original aim.

The embryo is in this context a biochemical system evolving from a single fertilized egg cell through a sequence of successive mitoses. First, two then four then eight etc. seemingly identical cells make up the embryo but at a certain stage the cells begin to differentiate, boundaries appear inside the system, and on the two sides of the boundaries tissues develop in different directions. It is through this process that different organs, bones, muscles, epidermis, etc. come into being and differentiate from each other. The problem is how we can grasp mathematically the process of differentiation which involves discontinuities in living tissue. The theory put forward by Thom in trying to solve this problem follows. We must note in advance that although the theory cannot claim much practical success, it still gives a valid framework of thoughts regarding this problem.

In an evolving embryo properties of living tissue are characterized by a very large amount of data, including values of the *intrinsic parameters* (or the *state variables*) of a system such as concentrations of different biochemical substances, the temperature, etc. The variation of these values in time can, in principle, be modeled by a continuous dynamical system, that is, by an autonomous system of differential equations of high dimension. The values of the intrinsic parameters, the densities, the temperature etc. may vary on a fast time scale and may tend to an attractor of the dynamical system. However, the dynamical system itself, which describes the biochemical dynamics, varies from point to point in the living tissue and also as time passes by on a slow time scale on which we observe the evolution of the embryo as a whole. This means that our dynamical system depends on four *external parameters*, that is, three spatial coordinates and "slow time." The dynamical system is supposed to be *smooth* (the right-hand side of the system of differential equations belongs to the C^∞ class as a function of the intrinsic and the external parameters). The vector of the state variables will be denoted by $x = (x_1, x_2, ..., x_n)$, the three spatial coordinates and the "slow time" will be denoted by $u_{1,}u_2, u_3$ and u_4, respectively, so that the vector of the external parameters is $u = (u_{1,}u_2, u_3, u_4)$. Thus, the evolution of the embryo is represented by the smooth autonomous system of differential equations

$$\dot{x} = f(x, u) , \tag{5.1.1}$$

where the overdot denotes derivative with respect to "fast time" t. At a certain point $(u_{1,}u_2, u_3)$ in tissue that, for example, will develop into the liver, and at a certain moment u_4 in the time interval during which the embryo becomes ripe to be born, system (5.1.1) normally has a single attractive equilibrium point and will settle down there. If u is varied just slightly, generically, the corresponding system behaves similarly, meaning it is equivalent to the neighboring ones, and has a similar attractor; in other words, by varying the external parameters continuously one expects that the attractors of the system vary continuously but their qualitative characteristics do not change. For these concepts see Appendix 2.3. However, it may happen that if certain points, curves, or surfaces are crossed in the space of the external parameters u the qualitative behavior of the attractors, their number or their stability changes, or they jump from one place to another. Where this happens, that is, where an arbitrary small variation of the external parameters may cause a qualitative change in the behavior of the system, is called the *bifurcation* or *catastrophe sets* of the system. Up to this point this is a bifurcation problem (treated in Appendix 2.3) that can be dealt with when the dynamical system is known. However, it was obvious when the theory was put forward (and it is yet the case) that we were unable to write out the system that modeled the embryo. Therefore, the problem had to be attacked from another direction. *The aim of the catastrophe theory was to classify all the possible qualitatively different ways how a system, depending on external parameters, may undergo a change in its qualitative behavior.* If such a classification were possible, then on the basis of observations one could tell what is happening at a certain bodily location on an embryo at a certain moment when the tissues differentiate. Unfortunately, this aim also turned out to be impossible to achieve a general form. However, based on in-depth results of topological dynamics the finite classification could be accomplished in the special case of *gradient systems* with four (later five) external parameters. This is now called the *elementary catastrophe theory.* We shall sum up its essence and the classification theorem for such systems. For more details see, for example, Poston and Stewart (1978).

System (5.1.1) is a *gradient system* if the vector field on the right-hand side is the negative gradient of a smooth scalar function $V : \mathbb{R}^n \times \mathbb{R}^r \longmapsto \mathbb{R}$, that is, $f = -\text{grad}_x V$, where the gradient is to be taken with respect to the first vector variable $x \in \mathbb{R}^n$, $\text{grad}_x V = [V'_{x_1}, V'_{x_2}, ..., V'_{x_n}]$, and smooth means of class C^∞. For the gradient system

$$\dot{x} = -\text{grad}_x V(x, u) \,, \tag{5.1.2}$$

the classification problem of vector fields depending on parameters is reduced to the classification of their "*velocity potential function*" V, which is somewhat easier. A gradient system has neither periodic (closed) nor homoclinic orbits (see Appendix 2.3 and Farkas, 1994). The motion governed by a gradient system is always in the direction of decreasing potential so that the minima of the potential are normally the asymptotically stable equilibria of the system (cf. Appendix 2.2).

We say that the point $(x, u) \in \mathbb{R}^n \times \mathbb{R}^r$ is a *critical point* of V if the gradient is zero at this point: $\text{grad}_x V(x, u) = 0$. The projections of the critical points to the space $\mathbb{R}^n$ are clearly the equilibria of the gradient system (5.1.2). The critical point is said to be *nondegenerate* or *hyperbolic* if at this point the Hessian of the function, that is, the determinant made up by the second derivatives is not zero, $\det[V''_{x_i x_k}] \neq 0$. We say that for a fixed $u \in \mathbb{R}^r$ the function $V(\bullet, u)$ is a *Morse function* if in every compact subset of $\mathbb{R}^n$ it has but a finite number of critical points and all the critical points are hyperbolic. Morse functions form a *generic* set in the space of all smooth functions from $\mathbb{R}^n$ to $\mathbb{R}$. This means that they form an open and dense set. *Open* means that every Morse function has a neighborhood in the space of functions in which all functions are Morse. *Dense* means that in an arbitrary small neighborhood of any function there is a Morse function. In other words, this means that if a Morse function is varied "slightly," it remains a Morse function, and that any smooth function may be moved into a Morse function by arbitrary small variation. What is "small" and what is "neighborhood" in the space of C^∞ functions is not easy to say. Roughly described, a function is near to another function if the modulus of their difference and the moduli of the differences of all their respective derivatives are small.

We turn first to the structural stability problem (cf. Appendix 2.3) of Morse functions and then to the same problem for families of functions depending on external parameters. If $a \in \mathbb{R}^n$ is a hyperbolic critical point of a function $F : \mathbb{R}^n \longmapsto \mathbb{R}, x \longmapsto F(x)$, then by *Morse's lemma* there is a smooth coordinate transformation or, in other words, a C^∞ diffeomorphism $h : \mathbb{R}^n \longmapsto \mathbb{R}^n, y \longmapsto h(y) = x$ such that $h(0) = a$ and in the new coordinates in a neighborhood of the point $y = 0$ the function assumes the form

$$(F \circ h)(y) = F(a) - y_1^2 - y_2^2 - \ldots - y_m^2 + y_{m+1}^2 + \ldots + y_n^2 \,. \tag{5.1.3}$$

The nonnegative integer $0 \leq m \leq n$ characterizes the critical point; if $m = 0$ then a is a minimum, if $m = n$ then it is a maximum point, if $0 < m < n$ then it is called a *Morse m-saddle.* The value m is an *invariant* of the function, it determines the *type* of the critical point; for any admissible coordinate transformation that carries the function F into the form (5.1.3) the value of m is the same. The type of a hyperbolic critical point is *structurally stable.* This means that any function close to F has a critical point of type m near to point a. It is to be noted that a hyperbolic critical point is always isolated, that is, it has a neighborhood in which there is no other critical point. We say also that the *function F is structurally stable in a neighborhood of its hyperbolic critical point.* Clearly, Morse functions are structurally stable at every critical point because every critical point is hyperbolic. As we see, here we have a generic set of structurally stable functions, and exactly this property makes the difference, that is, makes gradient systems more manageable than general smooth systems. As mentioned in Appendix 2.3, in the space of general systems of dimension higher than two, structural stability is *not* a generic property.

Let $F : \mathbb{R}^n \longmapsto \mathbb{R}, x \longmapsto F(x)$ be a smooth function with a critical point at a and $G : \mathbb{R}^n \longmapsto \mathbb{R}, x \longmapsto G(x)$ another smooth function with a critical point at b . We say that *F at a is equivalent to G at b* if there is a diffeomorphism (a smooth and smoothly invertible transformation) $h : \mathbb{R}^n \longmapsto \mathbb{R}^n$ and a real constant γ such that $h(b) = a$ and $(F \circ h)(x) + \gamma \equiv G(x)$ in a neighborhood of b. By applying this concept we may say that every Morse function at each critical point is equivalent to some quadratic function of the form $-y_1^2 - y_2^2 - \ldots - y_m^2 + y_{m+1}^2 + \ldots + y_n^2$ (at zero). It is easy to see that two Morse functions are equivalent to each other at corresponding critical points if and only if the critical points are of the same type.

If the smooth function $F : \mathbb{R}^n \longmapsto \mathbb{R}, x \longmapsto F(x)$ has a degenerate critical point at $a \in \mathbb{R}^n$, then the matrix $[F''_{x_i x_k}]$ at this point is singular; its rank is $< n$. For sake of convenience, the rank will be denoted by $n - q$ where $0 \leq q \leq n$ is called the *corank* of the matrix. (Corank zero means a regular matrix). The following theorem is the generalization of Morse's lemma to degenerate singular points.

THEOREM 5.1.1. *Suppose that F is a smooth function with a degenerate critical point at the origin and the corank of the matrix* $[F''_{x_i x_k}]$ *is* $q \geq 1$; *then*

F is equivalent to a function of the form

$$\widetilde{F}(x_1, x_2, ..., x_q) \pm x_{q+1}^2 \pm x_{q+2}^2 \pm ... \pm x_n^2$$

where $\widetilde{F}'_{x_i}(0) = 0$, $\widetilde{F}''_{x_i x_k}(0) = 0$, $(i, k = 1, 2, ..., q\ )$.

The assumption that the critical point is the origin does not restrict generality because, by a simple displacement of the origin into the critical point this situation can always be achieved. The importance of the theorem is that it reduces the study of the critical point to that of a function of fewer variables. The coordinates $x_1, x_2, ..., x_q$ are called the *essential* ones, and the function $\widetilde{F}$ is considered to be equivalent to F. This can be done because in the direction of the "nonessential" coordinates the critical point behaves like a Morse saddle. It can be proved that a degenerate critical point is never structurally stable, so that F, or equivalently, $\widetilde{F}$ in the theorem is not equivalent to every function in its neighborhood.

We shall extend these ideas to families of smooth functions dependent upon external parameters. The difficulty is that a family of functions dependent upon parameters may be structurally stable even if at certain values of the parameters the corresponding function of the family is *not* a Morse function. Nevertheless, the definitions of equivalence and structural stability are analogous to the single function case. Assume that we have two smooth families of functions $V : \mathbb{R}^n \times \mathbb{R}^r \longmapsto \mathbb{R}$, $(x, u) \longmapsto V(x, u)$ and $U : \mathbb{R}^n \times \mathbb{R}^r \longmapsto \mathbb{R}$, $(x, u) \longmapsto U(x, u)$; we say that they are *equivalent* (in a neighborhood of the origin, say) if a smooth function $h : \mathbb{R}^n \times \mathbb{R}^r \longmapsto \mathbb{R}^n$, a smooth function (a "constant" depending on the parameters u) $\gamma : \mathbb{R}^r \longmapsto \mathbb{R}$, and a diffeomorphism (a regular parameter transformation) $p : \mathbb{R}^r \longmapsto \mathbb{R}^r$ exist such that $h(0, u) = 0$, $h(\bullet, u)$ is a diffeomorphism for all u, $p(0) = 0$, and

$$V(h(x, u), p(u)) + \gamma(u) \equiv U(x, u)\ .$$

As before, the family V is said to be ***structurally stable*** if it is equivalent to every family sufficiently near to it. There is also a theorem analogous to Theorem 5.1.1 that makes it possible to reduce a function dependent upon external parameters to a function of the "essential" coordinates (and the parameters) only, which we shall not state here.

The main difficulty in presenting the elementary catastrophe theory lies in the problem of how many external parameters are needed for a structurally stable family of functions. If a function $V(x, u)$ where u are the external parameters is given it may happen, on the one hand, that some of the parameters may be discarded and what is left is still a structurally stable family, and on the other, that the family as it stands is not structurally stable but if more parameters are introduced it may be *embedded* into a structurally stable family. As there is no place here to present all the concepts needed and the exact theory, a shortcut is necessary. The usual procedure is to start from a degenerate critical point of a single function and construct a family of functions that contains the given one but is already structurally stable. When this is done the nonessential state variables (x coordinates) have already been eliminated by the analog of Theorem 5.1.1. The main result of Thom's Classification theorem is that it gives the number of different structurally stable families and characterizes them (up to four, later five external parameters). The theorem will be stated in a simplified form and for families of functions that have four external parameters at most, even though it also holds if this number is five. For the proof and the extension see Trotman and Zeeman (1976). The space of r parameter families of smooth functions of n state variables, that is, functions that map $\mathbb{R}^n \times \mathbb{R}^r$ into $\mathbb{R}$ will be denoted by $C^\infty(\mathbb{R}^n \times \mathbb{R}^r, \mathbb{R})$.

THEOREM 5.1.2. *If $r \leq 4$ then in the space $C^\infty(\mathbb{R}^n \times \mathbb{R}^r, \mathbb{R})$ there is a generic set $\mathcal{G}$ of families such that if $V \in \mathcal{G}$ and it is not a Morse family then after the reduction by the analog of Theorem 5.1.1 in a neighborhood of the origin it is equivalent to one of the following structurally stable families:*
if r=1: (i) to $x_1^3 + u_1x_1$, the "fold";
if r=2: (ii) to $\pm x_1^4 + u_2x_1^2 + u_1x_1$, the "cusp";
if r=3: (iii) either to $x_1^5 + u_3x_1^3 + u_2x_1^2 + u_1x_1$, the "swallowtail";
(iv) or to $x_1^3 + x_1x_2^2 + u_1x_1 + u_2x_2 + u_3x_1^2$, the "hyperbolic umbilic";
(v) or to $x_1^3 - x_1x_2^2 + u_1x_1 + u_2x_2 + u_3(x_1^2 + x_2^2)$,
the "elliptic umbilic";
if r=4: (vi) either to $\pm x_1^6 + u_4x_1^4 + u_3x_1^3 + u_2x_1^2 + u_1x_1$, the "butterfly";
(vii) or to $\pm(x_1^2x_2 + x_2^4) + u_1x_1 + u_2x_2 + u_3x_1^2 + u_4x_2^2$,
the "parabolic umbilic."

The seven families of the theorem are Thom's *elementary catastrophes.* In fact, their number is ten because in (ii), (vi) and (vii) the plus and minus signs yield different families; the "plus family" has minima where the "minus family" has maxima and vice versa. We have to warn the reader again that the theorem does not say that if one has a three-parameter family, for example, then it is necessarily equivalent to one of (iii), (iv), and (v). First of all, it may not belong to the set $\mathcal{G}$ of families, or it may belong to this set but it may contain a superfluous parameter, so that it may be equivalent to (ii). The expressions (i)-(vii) that occur in the theorem are the *unfoldings of the degenerate critical points* (or *singularities*) x_1^3, $\pm x_1^4$, x_1^5, $x_1^3{+}x_1x_2^2$, $x_1^3{-}x_1x_2^2$, $\pm x_1^6$, $\pm(x_1^2x_2{+}x_2^4)$, respectively. If a function with a degenerate critical point in the origin is given, in principle, one may determine the minimal number of parameters needed to produce a structurally stable unfolding, that is, a family that contains the given function and is structurally stable. This number can be any natural number or infinity. In the latter case there is no structurally stable unfolding. The theorem says that if this number is 1, 2, 3 or 4 then the stable unfolding of the function is necessarily equivalent to one of 1, 2, 3 or 4 different possibilities, respectively, given by the theorem. It is to be mentioned that there are also much finer classifications of singularities (see Arnold, 1975).

The simplest catastrophe on the list is the *fold.* The function $F(x) = x^3$ has a degenerate critical point in the origin because $F'(0) = F"(0) = 0$ (we drop the indices where possible in the sequel). Indeed, the functions of the unfolding $V(x, u) = x^3 + ux$ are arbitrary near to F in a sufficiently small neighborhood of the origin if $|u|$ is small, for $u \neq 0$ they are Morse functions; for $u > 0$ they have no critical points whatsoever and for $u < 0$ they have two hyperbolic critical points in the neighborhood of the origin, $x = \pm\sqrt{-u}$. The *family* V is structurally stable, containing, for $u = 0$, a structurally unstable (non-Morse) function. If we return to the corresponding gradient system (5.1.2) it is now $\dot{x} = -3x^2 - u$, presenting the saddle-node bifurcation at $u = 0$ described in Appendix 2.3.

The most popular item on the list is the *cusp*. This is so because it has richer properties than the fold and, at the same time, it can still be represented by 3D graphics. The function $F(x) = x^4$ has a degenerate minimum at the origin because $F^{'}(0) = F^{"}(0) = 0$. The "gradient" of the unfolding $V(x, u, v) = x^4 - ux^2 - vx$, where we chose negative signs for the sake of convenience, is $V_x^{'}(x, u, v) = 4x^3 - 2ux - v$ and the "Hessian" is $V_{xx}^{"}(x, u, v) = 12x^2 - 2u$. In this case $n = 1$ and $r = 2$. The critical points of the family lie on the surface $M = \{(x, u, v) \in \mathbb{R} \times \mathbb{R}^2 : 4x^3 - 2ux - v = 0\}$. They are degenerate if at the point in question the second derivative is zero, $12x^2 - 2u = 0$. The system of the two equations (the condition that the point should belong to M and that at the point the second derivative should be zero) can be solved easily for x and the solution yields the curve $\kappa = \{(u, v) \in \mathbb{R}^2 : v^2 = 8u^3/27\}$ in the plane of the parameters. This means that if $(u, v) \in \kappa$, then the corresponding element of the family has a degenerate critical point. The curve κ, called the *catastrophe set*, can be seen on Fig. 5.1.1.

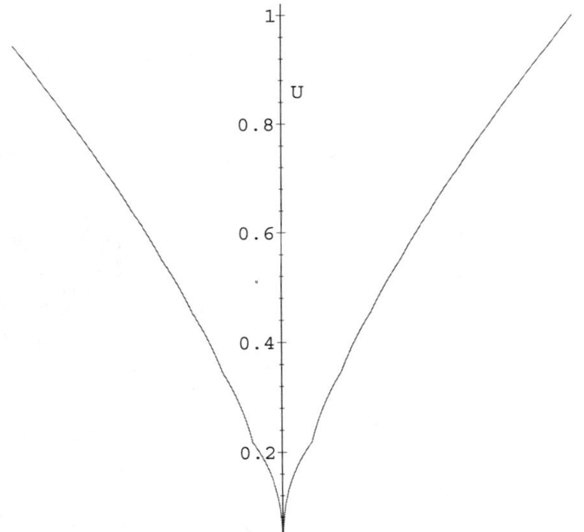

Figure 5.1.1: The catastrophe set of the cusp (MAPLE).

The form of the curve explains the name cusp. Simple calculations show that for parameter values outside the cusp the functions of the family have a single hyperbolic critical point, and if we look at the corresponding gradient system $\dot{x} = -4x^3 + 2ux + v$, this is asymptotically stable. For values of (u, v) inside the cusp there are three hyperbolic equilibria, with the greatest and the smallest asymptotically stable and the middle one unstable. On curve κ for $(u, v) \neq (0, 0)$ there are two equilibria, a fold-type degenerate (unstable) one and another hyperbolic (asymptotically stable) one. At $(u, v) = (0, 0)$ there is a single degenerate critical point that is asymptotically stable but not linearly stable (see Appendix 2.2). The surface M is shown on Fig. 5.1.2.

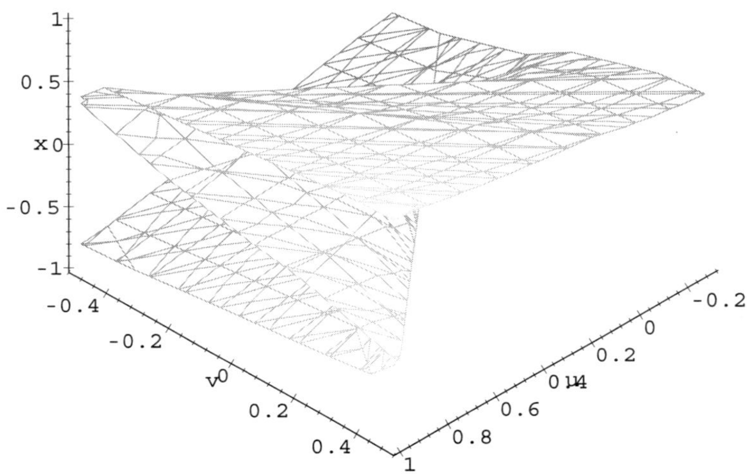

Figure 5.1.2: The surface M of critical points of the cusp (MAPLE).

The "catastrophic phenomena" occur if the parameters are continuously varied in the (u, v) plane. If for positive u coming from the left v is increased and the point (u, v) crosses the right branch of the catastrophe set κ the system that has been settled in a stable way at the attractor on the lower layer of the surface M loses this equilibrium and has to jump in a short time (the "fast time") to the only remaining attractor on the upper layer of M. If the point (u, v) is moving back on the same route to the left the system will not jump back to the old equilibrium at the place where it has left it but only when the route crosses the left branch of κ, that is, when the attractor on the upper layer has disappeared. If for $v = 0$ the value of u is increased from the back to the front and the point (u, v) crosses the cusp point, that is, the origin in the parameter plane, the system's qualitative properties change–before it had a single stable equilibrium, after the crossing it has two stable and one unstable equilibria (the pitchfork bifucation treated in Appendix 2.3 can be observed). What is important here is that these qualitative changes are consequences of the *continuous* variations of the external parameters.

A hypothesis put forward by Zeeman (1975) in the field of embryology will be

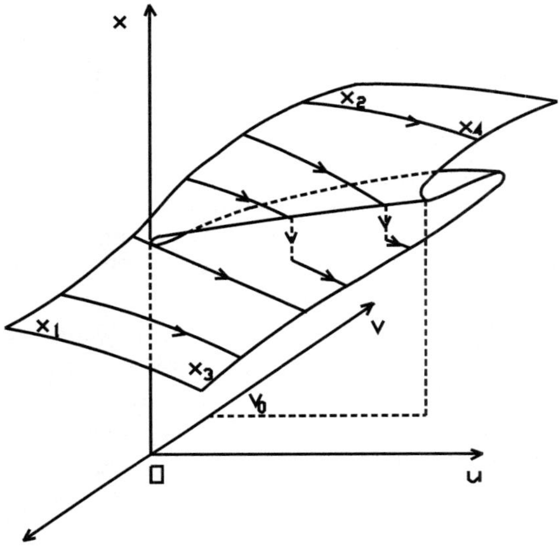

Figure 5.1.3: Differentiation of cells in embryonic tissues.

explained. Suppose a tissue in an embryo can be considered as a 1D ("linear") continuum so that the evolution of the cells in this tissue is influenced by two external parameters, a spatial coordinate v and the "slow time" u, and that the evolution of the cells is governed by a gradient dynamics, that is, the evolution is towards minimizing the value of a smooth potential. To be sure, we know neither the dynamics nor the potential function but if these assumptions are accepted then, qualitatively, the evolution can be modeled by the cusp catastrophe. We shall represent the state variables by a single coordinate x but, as explained earlier, this is not a real restriction and serves only the purpose of simpler drawings. Now, if we look at Fig. 5.1.2, cells which were at the beginning of the evolution to the left (on the spatial axis v), evolve towards the states represented by the lower layer of the surface M, and those that were to the right move towards states represented by the upper layer. While at the beginning ($u = 0$, say) their state varied only slowly and continuously on the axis v, at the end of the gestational period there will be a considerable discontinuity in their state somewhere in the middle of the tissue; the organs have differentiated in the process. However, the qualitative character of the cusp allows a lot of freedom in drawing the actual shape of M. The picture drawn by Zeeman looks somewhat like the one on Fig. 5.1.3. If it is valid the evolution is not so smooth. Cells to the far left evolve from state x_1 to state x_3, those to the far right evolve from the state x_2 to the state x_4, but those between the cusp point and the point v_0 start to develop towards state x_4 but having reached the edge jump down to the lower layer and will become x_3 cells. This "fast time" change may be the result of a triggering effect of some gene timed for some moment of the period of gestation. In fact, a wave moves along the tissue from the point $v = 0$ to the point $v = v_0$ while, finally, at the end of the process the frontier stabilizes at v_0. In the work quoted some phenomena concerning amphibia are shown which support this theory.

In Cooke and Zeeman (1976) the authors suggest a catastrophe theoretic model to describe a repeated pattern (such as vertebrate somite) formation.

The possibility of applying the elementary catastrophe theory is very much restricted by the condition that it concerns only gradient systems. This condition can be relaxed somewhat–the theory can be applied to dynamical systems that admit suitable Lyapunov functions (see Appendix 2.2), that is, scalar functions that decrease along the trajectories. In this case the study of the critical points of the Lyapunov function replaces that of the potential of the gradient system. Even if the theory can be applied one cannot expect exact quantitative results from it because it treats large classes of equivalent functions represented by a single function in normal form. The importance of the theory is that it yields a clear frame of thinking in many cases where exact models are not known.

5.2 Animal Hide and Shell Patterns

The fantastic variety of color and design observed in Nature of mammalian coats, tropical shells, butterflies, fish and the like has always attracted much human interest but attempts to provide a scientific explanation for this variety began in the second half of the twentieth century. The pattern of mammalian coat is determined by the genome of the given species and is formed at some stage of embryogenesis but the genes themselves cannot create the pattern. The first general theory based on the phenomenon of *Turing bifurcation* (see Appendix 3.3) and on Wolpert's positional information theory (see Wolpert 1981) was worked out by Murray and other authors (see Murray, 1989 and the references therein). In this Section we shall present this theory in some detail. At the end of the Section some other theories will also be mentioned with references.

According to *Murray's theory*, which is one of the so-called *"prepattern theories,"* the coat patterns of mammals (such as tigers, leopards, zebras etc.) are created in several steps during the evolution of the embryo. The first important step is the diffusion of a chemical *morphogen* over the surface of the developing embryo at an early stage of the gestational period. It is not known yet what this morphogen is but its spatial distribution over the integument of the embryo is the prepattern that determines the color pattern at a later stage. Much later in the gestational period certain cells (the would-be "melanocytes") wander and settle in the integument. These are able to produce melanin and create pigmentation provided that having read out their position from the concentration of the morphogen they are induced to do so and to differentiate (or, alternatively, they move on). The combined effect of the prepattern and the capability of certain cells to obtain *positional information* from chemical concentrations in their neighborhood and to produce melanin leads to the color patterns observed. In this complex process the characteristics of the different species are governed by the genetically determined prepattern, which is created by a reaction-diffusion process involving pattern formation due to Turing bifurcation. The assumption is that a substrate and a co-substrate react and diffuse in the would-be epidermis of the animal embryo and that these are certain enzymes that settle in a genetically determined pattern on a membrane and in a much later stage induce, activate, or inhibit melanin production. The integument is a 2D surface, considered in a rough approximation as either a rectangle or a kind of tubular surface, that is, a part of a cylinder or a cone. To illustrate the theory Murray used an experimentally tested model introduced by Thomas (1975) in another context. Denoting the density of the activator and the inhibitor at time t and place (x, y) (orthogonal Cartesian coordinates) by $u(t, x, y)$ and $v(t, x, y)$, respectively, the dynamics is supposed to be governed by the system of reaction diffusion equations

$$\frac{\partial u}{\partial t} = f(u, v) + \Delta u \ , \ \frac{\partial v}{\partial t} = g(u, v) + d\Delta v \ ,$$

where Δ is the "Laplace delta", $\Delta = \frac{\partial^2}{\partial x^2} + \frac{\partial^2}{\partial y^2}$ in Cartesian coordinates, $d > 1$ is the diffusion coefficient of the inhibitor (that of the activator has been transformed to 1), and

$$f(u, v) = a - u - \rho uv/(1 + u + Ku^2), \ g(u, v) = \alpha(b - v) - \rho uv/(1 + u + Ku^2). \tag{5.2.1}$$

Here all the parameters are considered to be positive. If the last term was not present in the expressions for f and g then clearly u and v would settle at $u = a$ and $v = b$. The last term has an inhibiting effect on the production of both u and v, but if u is sufficiently large, as u is increasing it is decreasing (because u is quadratic in the denominator and only linear in the numerator) and tending to zero, so that its effect is small if u is large, while it is increasing with v linearly, that is, v can be identified as the inhibitor. If the domain where the reaction-diffusion process takes place is the rectangle

$$R = \{(x, y) : 0 \leq x \leq p, 0 \leq y \leq q\}$$

and no-flux (Neumann) boundary conditions are assumed, meaning that no reactant is arriving to R from outside or leaving it, then these conditions are expressed by

$$u'_x(t, 0, y) = u'_x(t, p, y) = u'_y(t, x, 0) = u'_y(t, x, q) = 0$$

and identically for v. The results of the pattern formation process depend very much on the scale where it takes place, in other words, on how large the domain R is. It is therefore desirable to have the scale somehow present in the differential equations as a parameter. This can be achieved by transforming the coordinates and the time by

$$\tilde{x} = x/\sqrt{A} \ , \qquad \tilde{y} = y/\sqrt{A} \ , \qquad \tilde{t} = t/A \ , \tag{5.2.2}$$

where $A = pq$ is the area of the rectangle R. This transformation carries the system into

$$\tilde{u}'_{\tilde{t}} = A\, f(\tilde{u}, \tilde{v}) + \Delta \tilde{u}\,, \qquad \tilde{v}'_{\tilde{y}} = A\, g(\tilde{u}, \tilde{v}) + d\Delta \tilde{v} \tag{5.2.3}$$

where now $\Delta = \frac{\partial^2}{\partial \tilde{x}^2} + \frac{\partial^2}{\partial \tilde{y}^2}$, $\tilde{u}(\tilde{t}, \tilde{x}, \tilde{y}) = u(\tilde{t}A, \tilde{x}\sqrt{A}, \tilde{y}\sqrt{A})$ and similarly for v. The boundary conditions are also transformed, into

$$\tilde{u}'_{\tilde{x}}(\tilde{t}, 0, \tilde{y}) = \tilde{u}'_{\tilde{x}}\left(\tilde{t}, \sqrt{p/q}, \tilde{y}\right) = \tilde{u}'_{\tilde{y}}(\tilde{t}, \tilde{x}, 0) = \tilde{u}'_{\tilde{y}}(\tilde{t}, \tilde{x}, \sqrt{q/p}) = 0 \tag{5.2.4}$$

and identically for v. Observe that if the domain R can be considered a square, that is, $p = q$, then in the transformation (5.2.2) $\sqrt{A} = p$, the rectangle R goes into the square of side 1 and, accordingly, the boundary conditions also become simpler because $p/q = 1$. For convenience we drop the tilde in the notations in the sequel, so that in system (5.2.3) functions f, g are given by (5.2.1).

Now, the *kinetic system* belonging to (5.2.3) is

$$\dot{u} = A\, f(u, v)\,, \qquad \dot{v} = A\, g(u, v)\,. \tag{5.2.5}$$

First, we shall determine the positive equilibria of system (5.2.5), which are, at the same time, the constant stationary solutions of the problem (5.2.3)-(5.2.4). The equations of the $\dot{u} = 0$ and the $\dot{v} = 0$ isoclines of system (5.2.5) are

$$v = \left(a/u + a - 1 + (aK - 1)u - Ku^2\right)/\rho\,,$$

and

$$v = \alpha b(1 + u + Ku^2)/\left(\alpha + (\alpha + \rho)u + \alpha K u^2\right)\,,$$

respectively. To be able to perform the calculations we fix the values of the parameters numerically, $\alpha = 1.5$, $K = 0.1$, $\rho = 18.5$, $a = 92$, $b = 56$. Numerical calculations yield two equilibria with positive coordinates: $(u_1, v_1) = (3.958, 6.6025)$ and $(u_2, v_2) = (26.210, 12.813)$ (see Fig. 5.2.1). We are going to treat the second one. The derivative matrix of the right-hand side of system (5.2.5) is

$$-\frac{A}{1 + u + Ku^2}\begin{bmatrix} \frac{(1+u+Ku^2)^2 + \rho v(1-Ku^2)}{1+u+Ku^2} & \rho u \\ \rho v \frac{1-Ku^2}{1+u+Ku^2} & \alpha(1 + u + Ku^2) + \rho u \end{bmatrix}.$$

At (u_2, v_2) it is

$$A\begin{bmatrix} 0.7446 & -5.0556 \\ 1.7446 & -6.5556 \end{bmatrix}.$$

Calculating the trace and the determinant of this matrix we deduce that (u_2, v_2) is asymptotically stable (see Appendix 2.2). By linearizing the reaction-diffusion

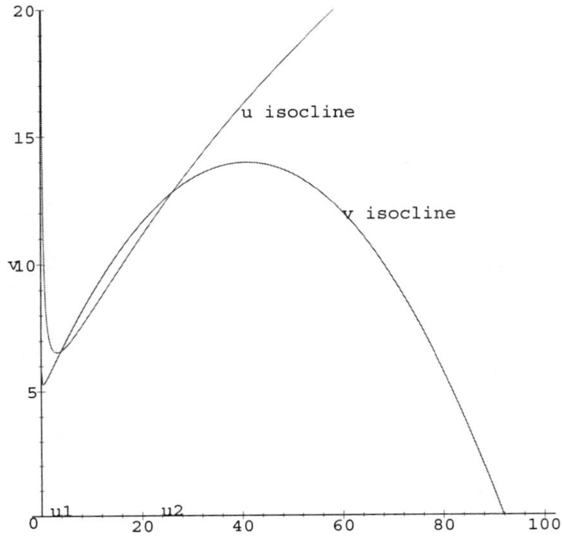

Figure 5.2.1: The zero isoclines and the equilibria of the kinetic system (5.2.5) (MAPLE).

system (5.2.3) at this equilibrium point and introducing the coordinates $U = u - u_2$, $V = v - v_2$, we obtain

$$\frac{\partial U}{\partial t} = A(0.7446U - 5.0556V) + \Delta U \qquad (5.2.6)$$
$$\frac{\partial V}{\partial t} = A(1.7446U - 6.5556V) + d\Delta V$$

with boundary conditions

$$U'_x(t,0,y) = U'_x(t,\sqrt{p/q},y) = U'_y(t,x,0) = U'_y(t,x,\sqrt{q/p}) = 0 \qquad (5.2.7)$$

and identically for V. Now we are to solve the linearized problem (5.2.6)-(5.2.7) by Fourier's method (the separation of the variables). The solution is similar to one shown in Appendix 3.3 in the general case but, where necessary, numerical values will be fixed more or less at will in order to obtain clear explicit results. According to the classical method applied, the solution satisfying system (5.2.6) and the homogeneous boundary conditions (5.2.7) is assumed in the form

$$\mathrm{col}[U(t,x,y), V(t,x,y)] = W(x,y)T(t) \ ,$$

where the 2D vector $T(t) = \mathrm{col}[T_1(t), T_2(t)]$ depends on the time only and the scalar function $W(x,y)$ depends on the spatial coordinates; thus the latter is to be determined so that the boundary conditions be satisfied. Substituting function WT into the system and dividing by W, the system

$$\begin{aligned}\dot{T}_1 &= A(0.7446T_1 - 5.0556T_2) + T_1\Delta W/W \\ \dot{T}_2 &= A(1.7446T_1 - 6.5556T_2) + dT_2\Delta W/W\end{aligned}$$

is obtained. If we imagine the first equation divided by T_1 and the second by T_2, all the terms depend on t except the last ones, which depend but on x, y. This may be so only if $\Delta W/W$ is a constant, that is, $\Delta W/W = -\lambda$. This way we obtain a 2D system of ODEs for T:

$$\dot{T} = (AF' - \lambda D)T \tag{5.2.8}$$

where

$$F' = \begin{bmatrix} 0.7446 & -5.0556 \\ 1.7446 & -6.5556 \end{bmatrix}, \quad D = \begin{bmatrix} 1 & 0 \\ 0 & d \end{bmatrix},$$

and an elliptic PDE for W

$$\Delta W = -\lambda W \tag{5.2.9}$$

with boundary conditions similar to (5.2.7). Eq. (5.2.9) is solved again by the separation of the variables, and W is assumed in the form $X(x)Y(y)$ as substituted into the differential equation, which is divided by XY to yield

$$Y''/Y = -\lambda - X''/X = -\nu\,,$$

a constant because the left-hand side depends only on y and the right-hand side only on x. In solving the differential equation $Y'' = -\nu Y$ it is easy to show that the solution satisfying the boundary conditions at $y = 0$ and $y = \sqrt{q/p}$ is $Y(y) = \cos(y\sqrt{\nu})$, provided that $\nu = k^2\pi^2 p/q$, $k = 0, 1, 2, \ldots$. Solving the differential equation $X'' = -(\lambda - k^2\pi^2 p/q)X$, the solution that satisfies the boundary conditions at $x = 0$ and $x = \sqrt{p/q}$ is $X(x) = \cos\left(x\sqrt{\lambda - k^2\pi^2 p/q}\right)$, provided that

$$\lambda = \lambda_{jk} = \pi^2(j^2 q/p + k^2 p/q)\,,\ j, k = 0, 1, 2, \ldots\,. \tag{5.2.10}$$

These are the *eigenvalues* of the problem (5.2.9) with the given boundary conditions. The corresponding solutions, the *eigenfunctions*, are

$$W_{jk}(x, y) = \cos\left(xj\pi\sqrt{q/p}\right)\cos\left(yk\pi\sqrt{p/q}\right).$$

By substituting the eigenvalue $\lambda = \lambda_{jk}$ into system (5.2.8), the latter has two eigenvalues μ_{jk1}, μ_{jk2}. The real parts of these are both negative if

$$\mathrm{Trace}(AF' - \lambda_{jk}D) < 0$$

and

$$\det(AF' - \lambda_{jk}D) > 0\,. \tag{5.2.11}$$

The trace is negative for every j, k because as already mentioned, the trace of the matrix AF' is already negative. If $\mu_{jk1} = \mu_{jk2}$, then clearly (5.2.11) holds too, so that if we look for a Turing bifurcation, that is, for the instability of the system (5.2.8), we may assume that the two μs are different. In this case the general solution of system (5.2.8) is

$$T^{jk}(t) = s^{jk1}\exp(\mu_{jk1}t) + s^{jk2}\exp(\mu_{jk2}t) ,$$

where s^{jk1} and s^{jk2} are eigenvectors corresponding to μ_{jk1} and μ_{jk2}, respectively. "The general solution" of the problem (5.2.6)-(5.2.7) is then

$$\begin{aligned} \begin{bmatrix} U(t,x,y) \\ V(t,x,y) \end{bmatrix} &= \sum_{j,k=0}^{\infty} (s^{jk1}\exp(\mu_{jk1}t) \\ &\quad + s^{jk2}\exp(\mu_{jk2}t)) \cos\left(xj\pi\sqrt{q/p}\right) \cos\left(yk\pi\sqrt{p/q}\right) . \end{aligned} \tag{5.2.12}$$

If an initial condition is prescribed, that is, $col[U(0,x,y), V(0,x,y)]$ is given, these functions are to be expanded into a Fourier series by the functions $\cos\left(xj\pi\sqrt{q/p}\right)\cos\left(yk\pi\sqrt{p/q}\right)$, and the coefficients of these expansions determine the (length of the) 2D coefficient vectors s^{jk1} and s^{jk2}. If (5.2.11) holds for all j, k then all the terms in the series (5.2.12) tend to zero as t tends to infinity and this means that the zero solution of the linearized problem (5.2.6)-(5.2.7) is asymptotically stable. This also implies that the spatially constant solution (u_2, v_2) of the nonlinear problem (5.2.3)-(5.2.4) is locally asymptotically stable, meaning that solutions with initial values sufficiently near to (u_2, v_2) tend to the latter as t tends to infinity. If this is the case, the diffusion process of the morphogen leads to an even concentration distribution over the epidermis and no patterns are formed.

For pattern production some terms in the series (5.2.12) shall not tend to zero with increasing time. We may find nonnegative μs if for some j, k inequality (5.2.11) is reversed. The equation

$$\det(AF' - \lambda D) = d\lambda^2 + A(6.5556 - 0.7446d)\lambda + 3.9387A^2 = 0 \tag{5.2.13}$$

will be considered. If this quadratic equation has no real root or both roots are negative then for positive λs (5.2.11) always holds. If this equation has positive roots and some λ_{jk} fall in the interval between the roots then for these the determinant is negative. The roots of the equation are

$$\lambda_{1,2} = \left(0.7446d - 6.5556 \pm \sqrt{0.5544d^2 - 25.5174d + 42.9759}\right) A/(2d) .$$

We have real roots only if the expression below the square root is positive. The roots of the quadratic polynomial below the square root are $d_{1,2} = 1.7533; 44.2738$. Values of d below 1.7533 are out because then both roots are negative, so that d must be larger than 44.2738. We set $d = 50$, that is, we assume that the diffusion rate of the inhibitor is 50 times that of the activator. With this value of d the canonical form of Eq. (5.2.13) is

$$\lambda^2 - 0.6135A\lambda + 0.0788A^2 = 0 \tag{5.2.13'}$$

whose roots are: $\lambda_1 = 0.1831A$, $\lambda_2 = 0.4305A$. If an eigenvalue λ_{jk} is situated between these two roots then $\mu_{jk1} < 0 \leq \mu_{jk2}$, so that one of the exponentials in the corresponding term in series (5.2.12) does not tend to zero with increasing time t. Thus, if there are such eigenvalues then the zero solution of the linearized problem (5.2.6)-(5.2.7) is no longer asymptotically stable and as a consequence the spatially constant solution (u_2, v_2) of the original nonlinear problem (5.2.3)-(5.2.4) is unstable. For those values of j, k for which $\mu_{jk2} = 0$ the corresponding terms in series (5.2.12) are stationary (independent of t) but are nonconstant in the spatial variables x, y. This is what is called a *pattern*. One may prove (see what is said about this in Appendix 3.3) that in this case the nonlinear problem (5.2.3)-(5.2.4) also has a spatially nonconstant stationary solution near to that of the linearized one. If $\mu_{jk2} > 0$, then the corresponding term in series (5.2.12) tends to infinity with increasing time. In this case we may suppose that in the nonlinear system some damping effect acts to keep the corresponding solution bounded. Although there is no mathematical proof for this, computer simulations suggest that this is so.

In order to illustrate the situation we fix now the ratio of q and p: $q/p = 4$. Then $A = 4p^2$ and $\lambda_{jk} = \pi^2(4j^2 + k^2/4)$. The smallest positive λ_{jk} is $\lambda_{01} = \pi^2/4 = 2.4674$. If A, that is, p is small, then clearly no eigenvalue falls into the interval $[\lambda_1, \lambda_2]$. As A is increased the roots of Eq. (5.2.13') increase, their distance increases too and, as a consequence, more and more λ_{jk} get between them (while some are left behind, getting out). With our choice $q/p = 4$ the condition

$$0.1831A \leq \lambda_{jk} = \pi^2(j^2 q/p + k^2 p/q) \leq 0.4305A$$

turns into

$$0.2960p^2 \leq 16j^2 + k^2 \leq 0.6976p^2\,. \tag{5.2.14}$$

Table 5.2.1 shows the index pairs of eigenvalues that satisfy (5.2.14) for some values of p.

TABLE 5.2.1 Index pairs with corresponding terms not tending to zero in series (5.2.12)

p	Index pairs (j, k) satisfying (5.2.14)
1	NONE
2	NONE
3	(0,2)
4	(0,3)
5	(0,3) (0,4) (1,0) (1,1)
6	(0,4) (0,5) (1,0) (1,1) (1,2) (1,3)

For illustration $p = 5$ is chosen. In this case the polynomial (5.2.13′) turns into $\lambda^2 - 61.35\lambda + 788.0$. This parabola and the smallest eigenvalues λ_{jk} are shown on Fig. 5.2.2.

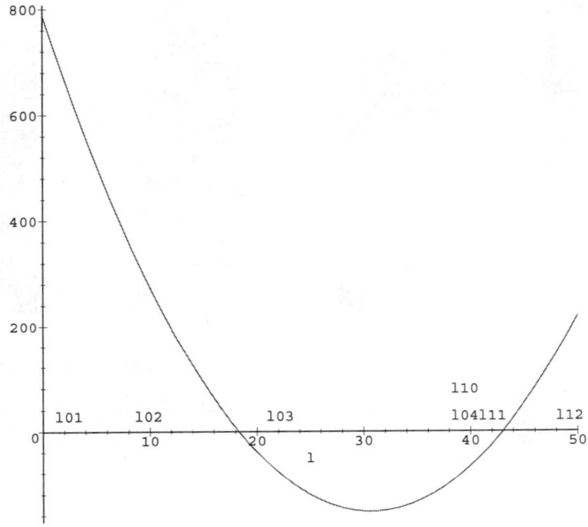

Figure 5.2.2: The parabola of Eq. (5.2.13′) with $p = 5$, that is, $A = 100$ and the smallest eigenvalues: $l_{01} = 2.5$, $l_{02} = 9.9$, $l_{03} = 22.2$, $l_{04} = 39.5$, $l_{10} = 39.5$ $l_{11} = 41.9$, $l_{12} = 49.3$ (MAPLE).

As seen on Fig. 5.2.2, in accordance with Table 5.2.1 those terms of (5.2.12) that do not tend to zero are

$$s^{032}\exp(\mu_{032}t)\cos(3\pi y/2), \quad s^{042}\exp(\mu_{042}t)\cos(2\pi y),$$
$$s^{102}\exp(\mu_{102}t)\cos(2\pi x), \quad s^{112}\exp(\mu_{112}t)\cos(2\pi x)\cos(\pi y/2).$$

According to what has already been said, we may suppose that the nonlinear problem (5.2.3)-(5.2.4) has a stationary solution near

$$\begin{bmatrix} u_2 \\ v_2 \end{bmatrix} + s^{032}\cos(3\pi y/2)$$

$$+s^{042}\cos(2\pi y) + s^{102}\cos(2\pi x) + s^{112}\cos(2\pi x)\cos(\pi y/2)\ ,$$

where the 2D vectors s^{jk2} have small coordinates. The pattern arises as these four modes get superposed upon the spatially constant equilibrial solution. On Fig. 5.2.3 the four modes are shown separately (assuming that the cosine terms have positive coefficients) on the $\sqrt{p/q} = 1/2$ times $\sqrt{q/p} = 2$ rectangle, with the dark regions representing higher and the white ones lower concentrations than the equilibrium.

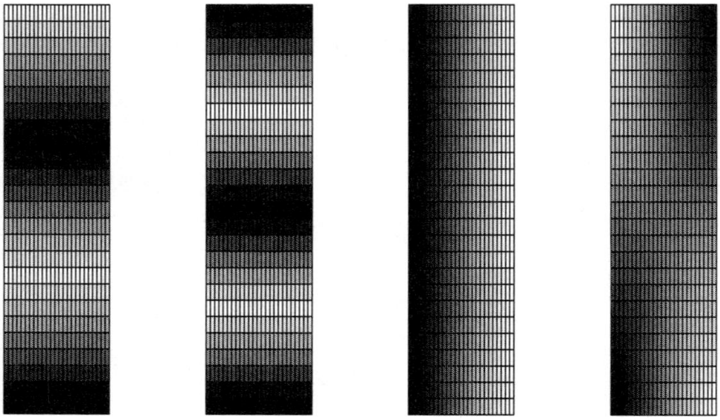

Figure 5.2.3: Patterns corresponding to the modes (0,3),(0,4),(1,0),(1,1) at $p = 5$ on the $\sqrt{p/q} = 1/2$ times $\sqrt{q/p} = 2$ rectangle in the x, y plane; higher and lower concentrations than at the equilibrium are represented by dark and white regions, respectively (MAPLE).

These simple calculations show how a great variety of patterns may arise with such a mechanism. At the same time this provides an explanation for the observed fact that small animals, such as mice and hares etc. and large ones, such as elephants, rhinoceroses and hippopotamuses do not have coat patterns. For small animals the domain of diffusion, that is, A in the previous example is small and therefore no eigenvalue falls into the interval of Turing instability. For very large animals too many eigenvalues fall into this interval and, as a consequence, too many modes get superposed upon each other. This results in a very fine pattern that is not recognizable.

The tail of a mammal can be considered as a very long cone that narrows as the end is approached. In the case of such a conical domain of diffusion one coordinate is measured along the length of the tail and the other one along the circumference of a cross section whose radius r is decreasing. For this second coordinate one has to prescribe "periodic boundary conditions" expressing the fact that point 0 of the circumference is identical to point $2r\pi$, so that at these points the concentrations must be equal. System (5.2.3) was solved numerically by Murray (1981) under such boundary conditions and the results coincided in a striking way with the observations. This can be seen on Fig. 5.2.4, which was used by Murray.

In recent years there has been intensive research on the "stripes vs spots" problem, that is, why one species has stripes and another of roughly the same size has spots. We have no place here to treat this problem and must be content with giving the references: Ermentrout (1991) and Lyons and Harrison (1991, 1992).

There are several different models that account for pattern formation in

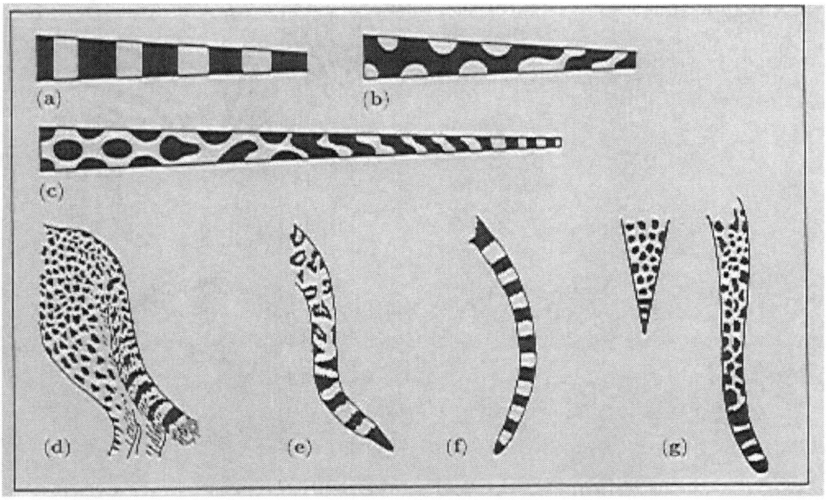

Figure 5.2.4: Computer simulation of mammal tail markings: (a),(b),(c); adult cheetah (*Acinonyx jubatis*) (d); adult jaguar (*Panthera onca*) (e); prenatal male genet (*Genetta genetta*) (f); adult leopard (*Felis pardus*) (g). Courtesy of Springer Verlag from Murray (1989) p.441.

living organisms (Murray, 1989) with applications to butterfly wing patterns, shell patterns, visual hallucination patterns etc.

Several activator-inhibitor models have been suggested by Meinhardt and Klinger (1987) (also see the beautifully illustrated book by Meinhardt, 1995) for the formation of tropical shell patterns. These are variations of the system

$$\begin{aligned} \frac{\partial u}{\partial t} &= s\left(u^2/v + b_u\right) - r_u u + D_u \frac{\partial^2 u}{\partial x^2} \\ \frac{\partial v}{\partial t} &= su^2 - r_v v + D_v \frac{\partial^2 v}{\partial x^2} + b_v \,, \end{aligned} \tag{5.2.15}$$

where u and v are the activator and the inhibitor, respectively, $b_u, b_v,\ r_u, r_v$, D_u, D_v are the respective basic production rates, decay rates, and diffusion coefficients, and s describes the ability of the cells to perform autocatalysis. Numerical solutions of the system produce many of the patterns observed in Nature. The growth of the shell and its shape also can be taken into consideration.

Appendix 1

DISCRETE MATHEMATICS

A1.1 Linear Algebra

Consider the n by n matrix $A = [a_{ik}]$ with real entries $a_{ik}, i, k = 1, 2, 3, \ldots, n$. We say that the nonzero vector $u = \text{col}\,[u_1, u_2, \ldots, u_n]$ (imagined as a column vector in matrix algebraic operations) with real or complex coordinates is an *eigenvector* corresponding to the real or complex *eigenvalue* λ if $Au = \lambda u$. The number λ is an eigenvalue iff it satisfies the *characteristic equation* $\det(A - \lambda I) = 0$ where I is the n-dimensional unit matrix. The *characteristic polynomial* $\det(A - \lambda I)$ is a polynomial of degree n and, as such, has n roots counting multiplicities. If λ_i and λ_k are two different eigenvalues then the corresponding eigenvectors are linearly independent. If all the eigenvalues are simple, that is, the characteristic polynomial has n different roots $\lambda_1, \lambda_2, \ldots, \lambda_n$, $\lambda_i \neq \lambda_k$ if $i \neq k$, then the corresponding eigenvectors $u^1, u^2, \ldots, u^n$ are linearly independent and form a basis of the n-dimensional space of column vectors; any vector can be expanded uniquely as a linear combination of these vectors. If a linear coordinate transformation is performed then the matrix is transformed by TAT^{-1} where T is the regular matrix of the transformation. Any n by n matrix B that can be written in the form $B = TAT^{-1}$ with a regular matrix T is said to be *similar* to A. Similar matrices have the same eigenvalues with the same multiplicities: the eigenvalues of a matrix with their multiplicities are *invariants* by regular coordinate transformation.

We say that the matrix is *stable* if the real parts of all eigenvalues are negative. We say that a polynomial is *stable* if the real parts of all its roots are negative. Thus, a matrix is stable if its characteristic polynomial is stable. There are ways to determine the stability of a polynomial without finding its roots. One of the most frequently used methods is called the *Routh-Hurwitz criterion*. It goes like this.

The polynomial is written in canonical form (the coefficient of the highest degree term is 1)

$$p_n(\lambda) = \lambda^n + a_{n-1}\lambda^{n-1} + a_{n-2}\lambda^{n-2} + \ldots + a_1\lambda + a_0 ,$$

where the a_k $(k = 0, 1, 2, \ldots, n-1)$ are real constants. The reader may prove it easily, just by examining the expressions for the roots, that a quadratic polynomial given in canonical form $p_2(\lambda) = \lambda^2 + a_1\lambda + a_0$ is stable iff both coefficients a_0, a_1 are positive. A polynomial of arbitrary positive degree can be decomposed over the real field into a product of first and second degree root-factors of the form $(\lambda - \lambda_k)$ and $(\lambda^2 + r_j\lambda + s_j)$. If the polynomial is stable then in every factor the coefficients $-\lambda_k, r_j, s_j$ are positive, so that by performing the multiplications and restoring the canonical form all coefficients will be positive. We arrived at a necessary condition of stability.

THEOREM A1.1.1. *If the polynomial $p_n(\lambda)$ given in canonical form is stable then every coefficient a_k, $(k = 0, 1, 2, \ldots, n-1)$ must be positive.*

In order to formulate a necessary and sufficient condition for the stability of a polynomial an n by n matrix, called the *Hurwitz matrix*, is formed by its coefficients:

$$H = \begin{bmatrix} a_1 & a_0 & 0 & 0 & 0 & 0 & \ldots & 0 \\ a_3 & a_2 & a_1 & a_0 & 0 & 0 & \ldots & 0 \\ a_5 & a_4 & a_3 & a_2 & a_1 & a_0 & \ldots & 0 \\ \cdot & \cdot & \cdot & \cdot & \cdot & \cdot & \ldots & \cdot \\ \cdot & \cdot & \cdot & \cdot & \cdot & \cdot & \ldots & \cdot \\ \cdot & \cdot & \cdot & \cdot & \cdot & \cdot & \ldots & \cdot \\ a_{2n-3} & a_{2n-4} & \cdot & \cdot & \cdot & \ldots & a_{n-1} & a_{n-2} \\ a_{2n-1} & a_{2n-2} & \cdot & \cdot & \cdot & \ldots & a_{n+1} & a_n \end{bmatrix}$$

where $a_n = 1$ and $a_k = 0$ for $k > n$.

THEOREM A1.1.2. *The polynomial $p_n(\lambda)$ given in canonical form is stable iff of its coefficients are positive, and all the principal diagonal minors are positive in its Hurwitz matrix.*

The proof can be found in Willems (1970).

For $n = 2$ the Hurwitz matrix is just

$$\begin{bmatrix} a_1 & a_0 \\ 0 & 1 \end{bmatrix},$$

so that the requirement of positivity of the principal diagonal minors does not impose any further condition on the stability of the polynomial beyond the positivity of the coefficients.

For $n = 3$ the Hurwitz matrix is

$$\begin{bmatrix} a_1 & a_0 & 0 \\ 1 & a_2 & a_1 \\ 0 & 0 & 1 \end{bmatrix},$$

so that beyond the positivity of the coefficients the necessary and sufficient condition of stability is

$$a_1 a_2 > a_0 . \tag{A1.1.1}$$

Often what is to be checked is not the stability of a polynomial but the question as to whether every root of the polynomial is in modulus < 1, or in other words, whether all the roots are in the interior of the unit circle around the origin of the complex plane. The Routh-Hurwitz criterion can be used for this purpose the following way. The transformation

$$\lambda = \frac{\nu + 1}{\nu - 1}$$

takes the interior of the unit circle of the complex plane λ into the interior of the left half plane of the complex plane ν. Thus, if we want to know whether all the roots of the polynomial

$$p_n(\lambda) = \lambda^n + a_{n-1}\lambda^{n-1} + a_{n-2}\lambda^{n-2} + \ldots + a_1\lambda + a_0$$

are in the interior of the unit circle we perform the transformation

$$P_n(\nu) = (\nu - 1)^n \, p_n\left(\frac{\nu + 1}{\nu - 1}\right) .$$

It is clear that P_n is also a polynomial of degree n, and that p_n has all its roots in the interior of the unit circle iff P_n is a stable polynomial. Applying the Routh-Hurwitz criterion to the latter polynomial for $n = 2$, and $n = 3$, respectively, we obtain the following criteria

The polynomial p_2 has both its roots in the interior of the unit circle iff

$$-1 + |a_1| < a_0 < 1 . \tag{A1.1.2}$$

The polynomial p_3 has all its roots in the interior of the unit circle iff

$$1 + a_1 > |a_0 + a_2| , \quad 3 - a_1 > |3a_0 - a_2| , \quad 1 - a_1 > a_0(a_0 + a_2) . \tag{A1.1.3}$$

In biological applications we are often confronted with matrices whose elements are positive or at least nonnegative. We say that a matrix (or a vector) is *positive* if all its entries are positive and we say that it is *nonnegative* if every entry is nonnegative. We say that a square matrix is *irreducible* if the same permutation of the rows and the columns cannot take it into a matrix that has but zeroes in a block in the upper right or in the lower left corner, that is, into a matrix that looks like

$$\begin{bmatrix} B & 0 \\ D & E \end{bmatrix} \quad \text{or} \quad \begin{bmatrix} B & C \\ 0 & E \end{bmatrix}$$

where B and E are square matrices. There holds the *Perron-Frobenius theorem* (see e.g., Marcus and Minc, 1964).

THEOREM A1.1.3. *If A is a nonnegative irreducible (respectively, positive) square matrix then it has a positive simple eigenvalue, every other eigenvalue is in modulus less than or equal to (respectively, definitely less than) this one, and this maximal positive eigenvalue has a positive eigenvector.*

In dealing with linear systems of differential equations we have to consider *complex analytic functions on matrices.* Let

$$f(\lambda) = \sum_{k=0}^{\infty} c_k \lambda^k$$

be a function analytic on an open disc around the origin of the complex plane λ, and A a square matrix whose eigenvalues lie in the interior of the disc of analiticity of the function f; then the square matrix $f(A)$ defined by

$$f(A) = \lim_{m\to\infty} \sum_{k=0}^{m} c_k A^k \tag{A1.1.4}$$

exists. If λ is an eigenvalue of A with u as the corresponding eigenvector then $f(\lambda)$ is an eigenvalue of $f(A)$ with eigenvector u, and $f(A)$ has no eigenvalues but of the form $f(\lambda)$ where λ is an eigenvalue of A. One may determine $f(A)$ without summing the infinite series that defines it (for details see, e.g., Farkas, 1994).

When biological phenomena are modeled by differential equations the values of the parameters involved can often be determined only crudely with significant errors. If we have a square matrix at hand it is important to know how much its stability depends on the actual values of the entries and how sensitive is it for variation of the entries. An important concept that can sometimes be applied is the concept of *sign stability* of a matrix.

DEFINITION A1.1.1. An n by n square matrix $A = [a_{ik}]$ is said to be *sign stable* if every n by n square matrix $B = [b_{ik}]$ of the same sign pattern, that is, such that sign b_{ik} = sign a_{ik} for all $i, k = 1, 2, \ldots, n$ is a stable matrix.

In other words this means that we may vary the entries of the matrix as long as positive, negative and zero entries stay positive, negative and zero, respectively, and thus the matrix remains stable.

There is an applicable criterion of sign stability due to Jeffries et al. (1977) that needs some graph theoretical preparation. To the n by n matrix $A = [a_{ik}]$ an undirected graph G_A is attached whose vertex set is V= $\{1, 2, \ldots, n\}$ and edges are $\{(i,j) : i \neq j,\ a_{ij} \neq 0 \neq a_{ji},\ \ i,j = 1,2,\ldots,n\}$. A directed graph D_A is also attached to A with the same vertex set and edges $\{(i,j) : i \neq j,\ a_{ij} \neq 0,\ \ i,j = 1,2,\ldots,n\}$. A *$k$-cycle* of D_A is a set of distinct edges of D_A of the form: $\{(i_1,i_2),(i_2,i_3),\ldots,(i_{k-1},i_k),(i_k,i_1)\}$. This is clearly equivalent to the condition that $a_{i_1i_2}a_{i_2i_3}\ldots a_{i_{k-1}i_k}a_{i_ki_1} \neq 0$. A subset of the vertices is $R_A = \{i : a_{ii} \neq 0\} \subset V$, those numbers for which the corresponding element in the main diagonal of the matrix is not zero. An *R_A-coloring* of G_A is a partition of its vertices into two sets, black and white (one of which may be empty), such that each vertex in R_A is black, no black vertex has precisely one white neighbor, and each white vertex has at least one white neighbor. A $V - R_A$ *complete matching* is a set M of pairwise disjoint edges of G_A such that the set of vertices of the edges in M contains every vertex in $V - R_A$ ($V - R_A$ is the set theoretical difference of V and R_A). By applying these concepts we are now able to state the following theorem, the proof of which can be found in Jeffries et al. (1977).

THEOREM A1.1.4. *An n by n real matrix $A = [a_{ik}]$ is sign stable iff it satisfies the following conditions:*

(i) $a_{ii} \leq 0$ *for all* i;

(ii) $a_{ij}a_{ji} \leq 0$ *for all* $i \neq j$;

(iii) *the directed graph* D_A *has no* k*-cycle for* $k \geq 3$;

(iv) *in every* R_A*-coloring of the undirected graph* G_A *all vertices are black; and*

(v) *the undirected graph* G_A *admits a* $V - R_A$ *complete matching.*

For example, the matrix

$$\begin{bmatrix} 0 & -1 & -1 & -1 \\ 1 & 0 & 0 & 0 \\ 1 & 0 & -1 & 0 \\ 1 & 0 & 0 & -1 \end{bmatrix}$$

satisfies all the conditions of the theorem and is sign stable.

There is also an only sufficient condition of sign stability

THEOREM A1.1.5. *If the matrix A satisfies conditions (i)-(iii) of the previous theorem and for each vertex i of G_A which belongs to one edge at most there holds $a_{ii} < 0$ then the matrix is sign stable.*

The previous example does not satisfy the conditions of the last theorem because $a_{22} = 0$. A matrix satisfying these conditions is

$$\begin{bmatrix} 0 & -1 & -1 & -1 \\ 1 & -1 & 0 & 0 \\ 1 & 0 & -1 & 0 \\ 1 & 0 & 0 & -1 \end{bmatrix}.$$

A1.2 One-Dimensional Discrete Semiflows

The set of natural numbers {0,1,2,3,...} will be denoted by $\mathbb{N}$, the set of integers by $\mathbb{Z} = \{\ldots, -2, -1, 0, 1, 2, \ldots\}$, the set of real numbers by $\mathbb{R}$, and the set of complex numbers by $\mathbb{C}$.

DEFINITION A1.2.1. Let $\varphi : \mathbb{N} \times \mathbb{R} \longmapsto \mathbb{R}$ be a real valued continuous function defined over the Cartesian product of the set of natural numbers and the set of real numbers; we say that φ is a *discrete forward dynamical system*, or a *discrete semiflow* if (i) $\varphi(0, x) = x$ for all $x \in \mathbb{R}$, (ii) $\varphi(k + l, x) = \varphi(k, \varphi(l, x))$ for all $k, l \in \mathbb{N}$, $x \in \mathbb{R}$.

The semiflow is called *discrete* because it is, in fact, a one-parameter family of continuous maps and the parameter varies in discrete steps over the set of natural numbers. It is *one-dimensional* because it acts on the set of real numbers. One can define similarly a semiflow over $\mathbb{R}^n$, in which case it is called n-dimensional. It is a *forward* dynamical system because if $\mathbb{N}$ is considered to be the model of "discrete time" then starting from a state x the semiflow φ takes it into the "future" $k > 0$ but cannot take it, necessarily, back into the "past." Condition (i) says that for $k = 0$ the map $\varphi(0, .)$ is the identity; condition (ii) says that if we start from a state x and reach a state in l steps and then start from this new state and reach a state in k steps, we arrive at the same position as if we had started from x and had moved forward $k + l$ steps. One may say, using the language of algebraic structures, that a semiflow is a one-parameter semigroup of continuous maps. One may define a semiflow on the Cartesian product of the set of natural numbers and an *interval* of the real line provided that φ takes the interval into itself for all $k \in \mathbb{N}$.

If the notation $F(x) = \varphi(1, x)$ is introduced, then from Definition A1.2.1, condition (ii), it is clear that $\varphi(2, x) = \varphi(1, \varphi(1, x)) = F(F(x)) = (F \circ F)(x)$ and

$$\varphi(k, x) = \varphi(1, \varphi(k-1, x)) = F\left(F^{k-1}(x)\right) = (F \circ F \circ \ldots \circ F)(x) = F^k(x)\ , \tag{A1.2.1}$$

that is, the semiflow is produced by the iterates of a continuous function $F : \mathbb{R} \longmapsto \mathbb{R}$. We say that the semiflow is *generated* by function F. The models considered in Section 1.1 are actually given by their generating functions, which is usually the case.

For any given x the set $\{\varphi(k, x) \mid k \in \mathbb{N}\}$ is called the *path (orbit, trajectory)* of x and the sequence $\varphi(k, x)$, $k \in \mathbb{N}$ is the *motion* along the path. We say that a point E is an *equilibrium point* or a *fixed point* of the semiflow φ if $\varphi(k, E) = E$ for all $k \in \mathbb{N}$. The path of an equilibrium point is a point path. A point P is said to be a *periodic point of period* n if $\varphi(n, P) = P$ and $\varphi(k, P) \neq P$ for any $k < n$. For the semiflow (A1.2.1) P is a periodic point of period n if it is a fixed point of the map F^n and it is *not* a fixed point of any F^k for $k < n$. The path of an n-periodic point P consists of exactly n points, namely, $P = P_0, P_1 = \varphi(1, P_0), P_2 = \varphi(2, P_0), \ldots, P_{n-1} = \varphi(n-1, P_0)$, and each point of this path is also an n-periodic point, as can be seen from Definition A1.2.1 (ii): $\varphi(n, P_k) = \varphi(n, \varphi(k, P)) = \varphi(n + k, P) = \varphi(k, \varphi(n, P)) = \varphi(k, P) = P_k$, for $k = 1, 2, \ldots, n-1$. We say that the subset H of $\mathbb{R}$ is an *invariant set* of the semiflow φ if $x \in H$ implies $\varphi(k, x) \in H$ for all $k \in \mathbb{N}$.

DEFINITION A1.2.2. The fixed point E of the semiflow φ is said to be *stable in the Lyapunov sense*, resp. *asymptotically stable*, resp. *unstable* if for any $\varepsilon > 0$ there exists a $\delta > 0$ such that $|x - E| < \delta$ implies $|\varphi(k, x) - E| < \varepsilon$, for all $k \in \mathbb{N}$, respectively, the previous condition holds and there exists a $\rho > 0$ such that $|x - E| < \rho$ implies $|\varphi(k, x) - E| \to 0$, as $k \to \infty$, respectively, it is not stable in the Lyapunov sense.

For semiflow (A1.2.1) it is easy to give sufficient conditions for stability.

THEOREM A1.2.1. *Assume that the function F is continuously differentiable and that E is an equilibrium point of the semiflow (A1.2.1), that is, $F(E) = E$; if $|F'(E)| < 1$ then E is asymptotically stable; if $F'(E) > 1$ then E is unstable.*

The proof of this theorem is straightforward with the Mean Value theorem applied. For details and more about what follows see Devaney (1989). Letting P be a periodic point of period n of the semiflow (A1.2.1), we say that P is *stable in the Lyapunov sense, asymptotically stable, unstable*, respectively, if it is a Lyapunov stable, asymptotically stable, unstable equilibrium, respectively, of the semiflow generated by function F^n. If F is continuously differentiable and the respective conditions of Theorem A1.2.1 hold for F^n at P then the path of P is also asymptotically stable or unstable in an obvious sense or, in other words, such are all the points $P = P_0, P_1, P_2, \ldots, P_{n-1}$ of the path. This is a consequence of the following Theorem.

THEOREM A1.2.2. *$F^{n'}(P) = F^{n'}(P_k)$ for all $k = 1, 2, \ldots, n-1$.*

PROOF. By the chain rule

$$\begin{aligned} F^{n'}(P) &= F'(P_{n-1}) F'(P_{n-2}) \ldots F'(P_0) \\ &= F'(P_{k-1}) F'(P_{k-2}) \ldots F'(P_0) F'(P_{n-1}) \ldots F'(P_k) = F^{n'}(P_k) \ . \end{aligned}$$

□

Thus, if $|F^{n'}(P)| < 1$ then P_k is also an asymptotically stable equilibrium point of the semiflow generated by F^n for $k = 1, 2, 3, \ldots, n-1$.

THEOREM A1.2.3. *If $|F^{n'}(P)| < 1$ then the motion of P is asymptotically stable in the following sense–if x is sufficiently close to P then the path of x stays close to the path of P, and $F^m(x) - F^m(P) \to 0$ as $m \to \infty$.*

PROOF. Because of continuity, if x is sufficiently close to P then $F^k(x)$ is sufficiently close to P_k, $k = 1, 2, \ldots, n-1$ and if $F^k(x)$ is sufficiently close to P_k then $F^{hn}(F^k(x)) - P_k \to 0$ as $h \to \infty$ (because of the asymptotic stability of P_k), but any natural number m can be written in the form $m = hn + k$ for some $k = 0, 1, 2, \ldots, n-1$ so that $F^{hn+k}(x) - F^{hn+k}(P) \to 0$ as $h \to \infty$ for $k = 0, 1, 2, \ldots, n-1$. □

The existence problem of an equilibrium for the semiflow (A1.2.1) can be solved, in principle, fairly easily. One has to solve the equation $x = F(x)$; graphically one has to find the point of intersection of the graph of the function F and the line $y = x$ in the Cartesian system of coordinates x, y (see Figs. 1.1.2 and 1.1.3). In order to find a 2-periodic path, say, one has to solve the equation $x = (F \circ F)(x)$, which may turn out to be much more difficult even in the case of a simple function F.

We are going to treat the problem as to how a 2-periodic path may turn up in a one parameter family of semiflows, each member of which has a fixed point. This problem leads us to the phenomenon of *period doubling bifurcations.* Bifurcation theory, in the context of dynamical systems, deals with systems depending on parameters and studies the problem of how qualitative properties of a system change as the parameters are varied. If some qualitative properties of a system change at a value of one of the parameters then we say that a *bifurcation* occurred, and the corresponding parameter value is called a *bifurcation point.*

Consider the one parameter family of real functions $F_\mu : \mathbb{R} \to \mathbb{R}$ where μ varies in an interval of the real line. The values of the function F_μ will also be denoted by $F_\mu(x) = F(\mu, x)$, and μ will be called the *bifurcation parameter.* We assume that F is a *continuous* function of its two variables and that for all μ the point $x = K$ is a fixed point, $F(\mu, K) = K$. This last assumption does not restrict generality provided that F_μ has a fixed point for all μ and it varies continuously with μ, because if $E(\mu)$ is this fixed point we introduce the new independent variable $z = K + x - E(\mu)$ and we arrive at a family with the required property. We attack the problem of how a 2-periodic path may arise around the equilibrium K in the family of semiflows generated by the family of functions F_μ as the bifurcation parameter is varied. When we say that the periodic path is around point K we mean that one of the points of the path is to the left of K and the other one is to the right of it. In simple generic cases this situation implies that F_μ is decreasing around K so that to the left of K its value is larger than K and to the right of K it is smaller (see Fig. A1.2.1 a). Figure A1.2.1 b also shows that in the case of the function $F(x) = -x + b$ every path is 2-periodic. The slope of this function is -1. It is easy to see by drawing graphs and can be proved by the Mean Value theorem that if the slope of a function around a fixed point is in modulus everywhere larger or everywhere smaller than -1, then there is no 2-periodic path in that neighborhood. Assume that F is continuously differentiable and that besides $F(\mu, K) \equiv K$ at the bifurcation point μ_0 "a 2-periodic path bifurcates out of K." This means that for values of μ arbitrarily near to μ_0 we have periodic points of period two in a neighborhood of K. Clearly, $F^2(\mu_0, K) - K = 0$, and if

$$\frac{\mathrm{d}}{\mathrm{d}x}\left(F^2(\mu, x) - x\right)_{(\mu_0, K)} = \left(F'_x(\mu_0, K)\right)^2 - 1 \neq 0\,,$$

then by the Implicit Function theorem there is a unique continuous function $x(\mu)$ such that $x(\mu_0) = K$ and $F^2(\mu, x(\mu)) - x(\mu) \equiv 0$. However, we know of such a function, $x(\mu) \equiv K$. As this function is unique we can not have nonconstant periodic trajectories of period two in a neighborhood of K. We see

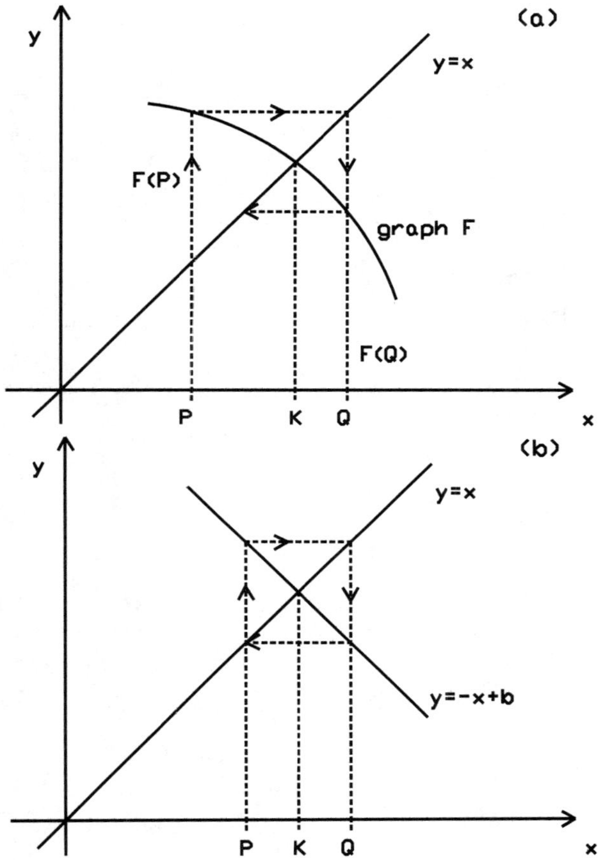

Figure A1.2.1: (a) The possibility of a 2-periodic path around the equilibrium K; and (b) for $F(x) = -x + b$ every path is 2-periodic.

that in order to have such trajectories we must have $(F'_x(\mu_0, K))^2 = 1$. But as F_μ must be decreasing $F'_x(\mu_0, K) \neq 1$, we must then have $F'_x(\mu_0, K) = -1$. This result motivates the following Theorem.

THEOREM A1.2.4. *Assume that the function F is twice continuously differentiable: (i) $F(\mu, K) - K \equiv 0$; and (ii) $F'_x(\mu_0, K) + 1 = 0$; and (iii) $F^{2''}_{x\mu}(\mu_0, K) \neq 0$; then there exists a neighborhood I of K and a unique continuously differentiable function $m : I \longmapsto R$ such that $m(K) = \mu_0$, $F(m(x), x) \neq x$, $x \neq K$ but $F^2(m(x), x) \equiv x$.*

The proof of this theorem can be found in Devaney (1989). The proof is not simple because the Implicit Function theorem can not be applied directly

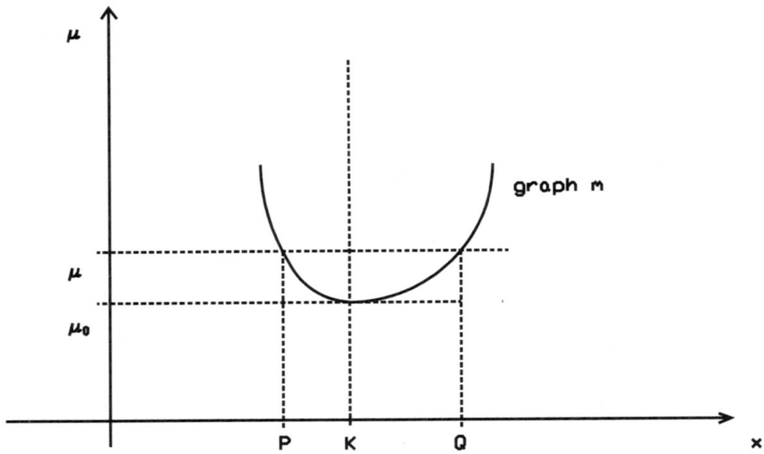

Figure A1.2.2: Bifurcation of a 2-periodic path P, Q from the equilibrium K at μ_0 for $\mu > \mu_0$.

to $F^2(\mu, x) - x = 0$. One can prove that under generic conditions $m'(K) = 0$ and $m''(K) \neq 0$. This means that the function m has either a minimum or a maximum at K, that is, the 2-periodic trajectories occur either for $\mu > \mu_0$ or for $\mu < \mu_0$. The graph of a typical function m with $m''(K) > 0$ is shown by Fig. A1.2.2. In many important cases we encounter the following phenomenon, where the fixed point K is asymptotically stable for $\mu < \mu_0$ by the inequality $F'_x(\mu, K) > -1$, at $\mu = \mu_0$ we have $F'_x(\mu_0, K) = -1$ by (ii) of the Theorem, for $\mu > \mu_0$ the equilibrium K is unstable (if $F^{2''}_{x\mu}(\mu_0, K) < 0$ at (iii) then, clearly, $F'_x(\mu, K) < -1$) and a periodic path of period 2 exists that is asymptotically stable.

In a large class of 1-parameter families of functions $F_\mu(x)$ one encounters the following situation, which is called *the period-doubling route to chaos*. For $\mu < \mu_0$ the function F_μ has an asymptotically stable fixed point K, at $\mu = \mu_0$ this equilibrium undergoes a period-doubling bifurcation and for $\mu_0 < \mu < \mu_1$ it is unstable and the system has an asymptotically stable periodic path P, Q. At $\mu = \mu_1$ both fixed points P and Q of the map F_μ^2 undergo a period doubling bifurcation, so that for $\mu_1 < \mu < \mu_2$ the 2-periodic path of F_μ becomes unstable and the twice two points of the two 2-periodic trajectories of F_μ^2 form an asymptotically stable periodic path of period 4 of F_μ, at $\mu = \mu_2$ this period 4 path gets destabilized and an asymptotically stable periodic path of period 8 bifurcates out of it, and so on. In this class of families an infinite sequence of period-doubling bifurcation occurs as the bifurcation parameter is increased and after all the periodic trajectories of period 2^k, $(k = 1, 2, 3, \ldots)$ have been destabilized a periodic trajectory of period 3 turns up. From a famous theorem by Sharkovskiy, if there is a periodic trajectory of period 3 in such a system then there are periodic trajectories of every period $1, 2, 3, \ldots$. When we arrive at this stage the dynamics of the semiflow becomes *chaotic*. A chaotic dynamics means three things. First, it is *sensitive to initial conditions*; this means that some motions that start arbitrarily near to each other will get farther from each other than a given positive number in the future. Second, it is *topologically transitive*, meaning that the image of any open interval will meet any other open interval some time in the future. Third, the *periodic points form a dense set* in the domain of the map that defines the semiflow.

An often quoted example of a family of semiflows that shows this phenomenon is the one given in Eq. (1.1.2). If a new variable is introduced by the transformation $x = Nr/((1+r)K)$, then the semiflow assumes the form

$$x_{k+1} = (1+r)\, x_k\, (1 - x_k)\ . \tag{A1.2.2}$$

It is easy to see by applying Theorem A1.2.1 that the equilibrium point $x = r/(1+r)$ of this system is asymptotically stable for $0 < r < 2$; at $r = 2$ this fixed point undergoes a period doubling bifurcation and as r is increased up to $r = 2.828\ldots$ the system goes through an infinite sequence of period doubling bifurcations. Having achieved this value the system has an infinite number of periodic solutions of periods 2^k $(k = 1, 2, 3, \ldots)$ all of which have become unstable. At this value of the bifurcation parameter r a periodic solution of period 3 turns up and if r is larger than this critical value the dynamics is chaotic. System (1.1.3) shows similar behavior. By the way, if we expand the per capita growth rate of system (1.1.3) by powers of $(1 - N/K)$ we obtain

$$e^{r(1-N/K)} - 1 = \sum r^k\, (1 - N/K)^k\, /k! \quad - 1 = r\,(1 - N/K) + \text{h.o.t.}\,,$$

(higher order terms) that is, the per capita growth rate of the logistic system is just the first approximation for that of the exponential one.

For system (1.1.3) the derivative of the generating function $F(N) = Ne^{r(1-N/K)}$ is

$$F'(N) = e^{r(1-N/K)}\,(1 - rN/K)\ .$$

At $N = 0$ we have $F'(0) = e^r > 1$, that is, the uninteresting equilibrium $N = 0$ is unstable, and $F'(K) = 1 - r$, implying that the equilibrium $N = K$ is asymptotically stable for $0 < r < 2$. At $r = 2$ this equilibrium undergoes a period-doubling bifurcation. This can be seen as follows. If P and Q are points of a 2-periodic path then they must satisfy the system of equations

$$Q = Pe^{r(1-P/K)}, \qquad P = Qe^{r(1-Q/K)},$$

that is, by substituting the equation

$$P = Pe^{r(1-P/K)}e^{r\left(1-(P/K)e^{r(1-P/K)}\right)}, \text{ or } \left(1 + e^{r(1-P/K)}\right) P/K = 2\,.$$

However, by the substituting $P = K(1+x)$ the last equation goes into

$$x = 2/\left(1 + e^{-rx}\right) - 1 = \left(1 - e^{-rx}\right) / \left(1 + e^{-rx}\right) = \tanh\left(rx/2\right) \tag{A1.2.3}$$

see Svirezhev-Logofet (1983). Now, the maximum of the slope of function $\tanh(rx/2)$ is assumed at $x = 0$ and it is $r/2$, so that the straight line $y = x$ intersects the graph of this function in two points besides $x = 0$ iff $r/2 > 1$, that is, $r > 2$. Under this last condition (A1.2.3) has exactly two nonzero solutions x_1, x_2, and as a consequence, $P = K\,(1+ x_1), Q = K(1 + x_2)$ form a 2-periodic path of the semiflow. This period 2 path stays asymptotically stable up to the value $r = 2.526$ of the bifurcation parameter. At this value it becomes unstable and an asymptotically stable period 4 path bifurcates out of it. As r is increased further the system passes through an infinite sequence of period-doubling bifurcations, and at $r = 3.102$ a periodic path of period 3 appears. Beyond this value the dynamics is chaotic. On Figs. 1.1.4, 1.1.5 and 1.1.6 the 4-periodic path, the 8-periodic path and the chaotic stage are shown at $r = 2.65, 2.67$, and 3.3, respectively.

A1.3 Discrete Dynamical Systems

We say that the function $F : \mathbb{R}^n \longmapsto \mathbb{R}^n$ is a *diffeomorphism* if it is one-to-one, that is, it is invertible and continuously differentiable (once, at least) along with its inverse function F^{-1}. If the function F is given by its coordinate functions

$$F(x) = (F_1(x_1, x_2, \ldots, x_n), F_2(x_1, x_2, \ldots, x_n), \ldots, F_n(x_1, x_2, \ldots, x_n)) \ ,$$

where x_k is the kth coordinate of x then the Jacobi (derivative) matrix of F will be denoted by

$$DF = \begin{bmatrix} F'_{1x_1} & F'_{1x_2} & \cdots & F'_{1x_n} \\ F'_{2x_1} & F'_{2x_2} & \cdots & F'_{2x_n} \\ \cdot & \cdot & \cdots & \cdot \\ F'_{nx_1} & F'_{nx_2} & \cdots & F'_{nx_n} \end{bmatrix} \ ; .$$

If F is continuously differentiable and $\det DF \neq 0$ then it is a diffeomorphism, at least locally, that is, in the neighborhood of any point. Diffeomorphisms also can be defined, naturally, over subsets of $\mathbb{R}^n$.

DEFINITION A1.3.1. Let $\varphi : \mathbb{Z} \times \mathbb{R}^n \longmapsto \mathbb{R}^n$ be a function defined over the Cartesian product of the set of integers and $\mathbb{R}^n$; we say that φ is a *discrete flow* (or *dynamical system*) if (i) for each $k \in \mathbb{Z}$ the function $\varphi(k, \bullet)$ denoted also by φ_k is a diffeomorphism; (ii) $\varphi(0, x) = x$ for all $x \in \mathbb{R}^n$, that is, φ_0 is the identity; (iii) for all $k, l \in \mathbb{Z}$ and $x \in \mathbb{R}^n : \varphi(k + l, x) = \varphi(k, \varphi(l, x))$.

If the notation $F(x) = \varphi(1, x)$ is introduced then, clearly, $\varphi(k, x) = F^k(x)$, $k = 1, 2, 3, \ldots$ where F^k is the kth iterate of F, also $F^0(x) \equiv x$, and F^{-1} is the inverse of F by (iii). This also implies that $\varphi(-k, x) = F^{-k}(x)$ is the inverse function of F^k, $k = 1, 2, 3, \ldots$. We say then that the function F *generates* the flow. A flow differs from a *semi* flow in that its generating function is invertible and, as a consequence, a flow may carry a point x of $\mathbb{R}^n$ back in time or, in other words, in case of a flow one may speak not only about the future but also about the past of a state x. A path, a motion, a fixed point, a periodic path, an invariant set, Lyapunov stability, and asymptotic stability are defined in an analogous way. We say that an invariant set in $\mathbb{R}^n$ is an *attractor* if it has a neighborhood such that for any point in that neighborhood the corresponding path tends to the invariant set as time k tends to infinity. We say that an invariant set is a *repellor* if it has a neighborhood such that the path of any point in that neighborhood tends to the set as time k tends to minus infinity (this means that the path tends away from the set as time is increasing). An asymptotically stable equilibrium or periodic path is obviously an attractor.

The analog of Theorem A1.2.1 about the stability of a fixed point is now:

THEOREM A1.3.1. *Assume that $E \in \mathbb{R}^n$ is an equilibrium point of the flow generated by F, that is, $F(E) = E$; if all the eigenvalues of the matrix $DF(E)$ are in modulus < 1 then E is asymptotically stable; if one of the eigenvalues is > 1 then E is unstable.*

For the proof of the Theorem see Devaney (1989).

Consider now a periodic path $P_0, P_1, P_2, \ldots, P_{m-1}$ of period m of the flow generated by the function F, that is, let $F(P_k) = P_{k+1}$, $P_m = P_0$, $(F^m(P_k) = P_k)$, for $k = 0, 1, 2, \ldots, m-1$. For the stability of the periodic path we have the following obvious theorem.

THEOREM A1.3.2. *If all the eigenvalues of the matrix $DF^m(P_0)$ are in modulus < 1 then the periodic path is asymptotically stable; if at least one eigenvalue is in modulus > 1 then it is unstable.*

We note here that now we cannot say that all the matrices $DF^m(P_k)$, $k = 0, 1, 2, \ldots, m-1$ are equal but it is fairly easy to show that they are *similar*, that is, their eigenvalues are the same, so that the stability criterion does not depend on the point of the periodic path where we try to apply it. For methods to check this stability criterion see Appendix 1.1.

Now we treat the stability problem of the equilibria of the flow (1.1.6). The Jacobi matrix of the pair of functions (F, G) is

$$J(N,P) = \begin{bmatrix} (1 - Nr/K)\, e^{r(1-N/K-Pa/r)} & -aNe^{r(1-N/K-Pa/r)} \\ 1 - e^{-aP} & aNe^{-aP} \end{bmatrix}.$$

Note that

$$\det J(N,P) = aNe^{r(1-N/K-Pa/r)}\left(1 - re^{-aP}N/K\right) > 0$$

provided that $0 < N < K$, $0 \leq P$ and $r \leq 1$, that is, in the domain of the N, P plane that interests us the map generated by the pair (F, G) is a local diffeomorphism indeed if r is < 1. If r is > 1 then at some points of this domain of the plane the map may not be one-to-one.

The characteristic equation of the matrix J at the fixed point $(0,0)$ is

$$\det(J(0,0)-\lambda I)=\lambda^2-e^r\lambda=0\,,$$

where I is the unit matrix. Because one of the roots is $e^r>1$, this equilibrium is unstable.

At the fixed point $(K,0)$ we have

$$\det(J(K,0)-\lambda I)=\lambda^2+\lambda\,(r-aK-1)+aK\,(1-r)=0\,.$$

If $r<aK+1$ then the stability condition (A1.1.2) is $aK<1$ and $aK(1-r)<1$; however, the first condition contradicts (1.1.9). If $r>aK+1$ then the conditions are $r<2$ and $aK(1-r)<1$; however, (1.1.9) implies that now $r>2$, which is again a contradiction. This means that the equilibrium $(K,0)$ is also unstable.

At the fixed point $E=\left(\bar{N},\bar{P}\right)$ in the interior of the positive quadrant of the plane the characteristic equation is

$$\det\left(J\left(\bar{N},\bar{P}\right)-\lambda I\right)=\lambda^2+\lambda\left(r\bar{N}/K-a\bar{N}e^{-r\left(1-\bar{N}/K\right)}-1\right)+$$

$$+a\bar{N}\left(1-re^{-r\left(1-\bar{N}/K\right)}\bar{N}/K\right)=0\,. \qquad \text{(A1.3.1)}$$

As the coordinates $\left(\bar{N},\bar{P}\right)$ can be determined only numerically the stability condition (A1.1.2) also is to be checked this way. Some results partly shown on Fig. 1.1.8 are displayed in Table A1.3.1.

TABLE A1.3.1

	K	a	r	E	Character
1	1	1.5	0.5	(0.71525, 0.09492)	stable, Fig.1.1.8a
2	1	1.5	1.5	(0.78174, 0.21826)	stable
3	1	1.5	2.5	(0.83065, 0.30482)	stable
4	1	1.5	3	(0.83973, 0.32054)	unstable
5	10	0.33320	1		unstable; attractive closed curve
6	10	0.42998	2		unstable; attractive closed curve, Fig.1.1.8b
7	10	0.48203	2.48800		unstable; attractive 20-periodic path Fig. 1.1.8c
8	10	0.51178	2.75		unstable; chaotic attractor, Fig. 1.1.8d

A1.4 Branching Processes

Relying only on some basic knowledge of probability theory, the concept of "generator function" will be introduced and then some more or less intuitively clear results will be established concerning "branching processes."

Suppose that ξ is a random variable that may assume only nonnegative integer values $0, 1, 2, 3, \ldots$ with probabilities $p_0, p_1, p_2, p_3, \ldots$, respectively, $p_i \geq 0$, $\sum_{i=0}^{\infty} p_i$ convergent.

DEFINITION A1.4.1. The function defined by

$$g(s) = \sum\nolimits_{k=0}^{\infty} p_k s^k$$

is called the *generator function* of the random variable ξ.

In words, the generator function is the sum of the power series in which the coefficient of the kth power of the variable is the probability of assuming the value k. It is clear that the function g is well defined for $s \in (-1,\ 1]$, and that it is analytic in the interior of this interval. The generator function defines the probability distribution uniquely. The mathematical expectation of ξ is

$$E(\xi) = g'(1) = \sum\nolimits_{k=1}^{\infty} p_k k \tag{A1.4.1}$$

provided that the last series is convergent. The property established in the following theorem will be used later.

THEOREM A1.4.1. *The generator function of the sum of independent random variables is equal to the product of their generator functions.*

PROOF. The theorem will be proved for the case of two random variables; the general case follows by mathematical induction. Let the generator functions of the random variables ξ and η be $g(s)$ and $h(s) = \sum_{k=0}^{\infty} q_k s^k$, respectively.

$$g(s)\, h(s) = \sum\nolimits_{k=0}^{\infty} \left(\sum\nolimits_{j=0}^{k} p_{k-j} q_j \right) s^k ,$$

but $\sum_{j=0}^{k} p_{k-j} q_j$ is obviously, the probability that the sum $\xi + \eta$ of the independent random variables assumes the value k. □

Consider now a population of a finite number of individuals who are reproducing independently by the same probability distribution. The probability of an individual having $\xi = 0, 1, 2, 3, \ldots, k, \ldots$ offspring in the next generation is $p_0, p_1, p_2, p_3, \ldots, p_k, \ldots$, respectively, $p_i \geq 0$, $\sum_{i=0}^{\infty} p_i = 1$. The generations are supposed to be discrete, and each individual has the same life span. Denoting the number in the 0th, 1st, 2nd, 3rd, etc. generations by $X_0, X_1, X_2, X_3, \ldots$ the sequence of the random variables X_n, $n = 0, 1, 2, \ldots$ is called a *branching process* (it is a special case of a "Markov chain"; for this and other details see Karlin-Taylor, 1975). We are interested first of all in the question, what is the probability of a population dying out, for example, in the nth generation? The conditional probability of $X_n = 0$, provided that $X_0 = k$ is, clearly, the kth power of the conditional probability of $X_n = 0$ provided that $X_0 = 1$:

$$P(X_n = 0 \,|X_0 = k) = (P(X_n = 0 \,|X_0 = 1))^k ,$$

because the descendants of each individual present at the beginning must die out independently of the rest of the population. Therefore, from now on we assume that $X_0 = 1$. Under this condition $\xi = X_1$. The generator function of ξ, and X_n, respectively, will be denoted by

$$g(s) = \sum_{k=0}^{\infty} p_k s^k , \; g_n(s) = \sum_{k=0}^{\infty} P(X_n = k) s^k , \text{ respectively.}$$

Clearly, $g_0(s) \equiv s$ and $g_1(s) \equiv g(s)$. A recursion formula will be deduced for g_{n+1},

$$\begin{aligned} g_{n+1}(s) &= \sum_{k=0}^{\infty} P(X_{n+1} = k) s^k = \\ &= \sum_{k=0}^{\infty} \sum_{j=0}^{\infty} P(X_{n+1} = k \,|X_n = j) P(X_n = j) s^k \\ &= \sum_{j=0}^{\infty} P(X_n = j) \sum_{k=0}^{\infty} P(X_{n+1} = k \,|X_n = j) s^k , \end{aligned}$$

where the Law of Total Probability was used. Now, the probability that in the $(n+1)$st generation we have k individuals provided that in the nth generation we had j is equal to the probability that those j individuals have altogether k offsprings. Denoting the random variable of the number of the offspring of the ith individual by ξ_i, $(i = 1, 2, \dots, j)$, these are independent random variables, each having the same distribution as ξ. Thus,

$$\sum_{k=0}^{\infty} P(X_{n+1} = k \,|X_n = j) s^k = \sum_{k=0}^{\infty} P\left(\sum_{i=1}^{j} \xi_i = k\right) s^k .$$

However, on the right-hand side we have the generator function of $\sum_{i=1}^{j} \xi_i$ which is, by Theorem A1.4.1, the jth power of g. Hence,

$$g_{n+1}(s) = \sum_{j=0}^{\infty} P(X_n = j) (g(s))^j .$$

On the right-hand side we have the generator function of X_n taken at $g(s)$, so that $g_{n+1}(s) = g_n(g(s))$ or $g_{n+1} = g_n \circ g$. Continuing the recursion we obtain that $g_{n+1} = g \circ g \circ \dots \circ g$ or

$$g_{n+1}(s) = g(g_n(s)) . \tag{A1.4.2}$$

We may now determine the mathematical expectation of X_n easily,

$$E(X_n) = g_n'(1) = (g'(1))^n = (E(\xi))^n \tag{A1.4.3}$$

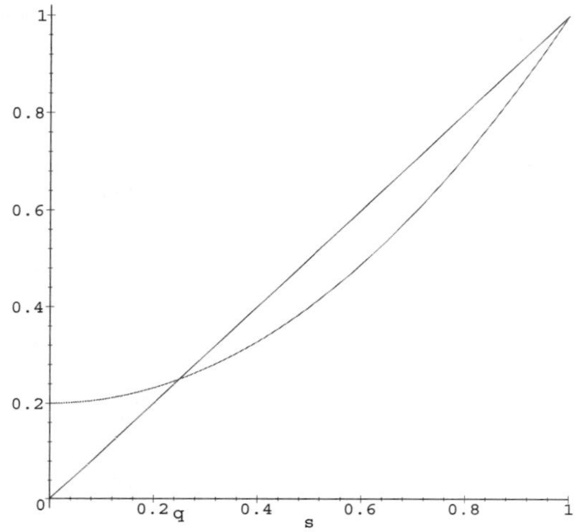

Figure A1.4.1 The graph of the generator function g in case the mathematical expectation of the number of offspring is > 1; the probability q of dying out.

where the chain rule, $g(1) = 1$ and Eq. (A1.4.1) were used.

Finally, we may attack the basic problem–what is the probability of the population dying out? The probability that the population dies out by the nth generation is $P(X_n = 0) = g_n(0)$. As n tends to infinity this is a nondecreasing sequence of nonnegative numbers ≤ 1. We call the limit of this sequence the probability of the dying out of the population and denote it by $q = \lim g_n(0)$. By Eq. (A1.4.2) this probability is the root of the equation

$$q = g(q) : . \tag{A1.4.4}$$

Suppose that $p_2 + p_3 + \ldots > 0$. Then g is increasing and convex down for $s > 0$, so that its graph may have two points of intersection with the 45° line at most. Let r be an *arbitrary* positive root of Eq. (A1.4.4). We shall show that for all n we have $g_n(0) \leq r$. Indeed, $g(0) < g(r) = r$, and assuming $g_n(0) < r$ one gets $g_{n+1}(0) = g(g_n(0)) \leq g(r) = r$, which proves the statement. This implies that the sequence $g_n(0)$ tends to the *smaller* root of Eq. (A1.4.4). A simple geometrical inspection shows that if $E(\xi) = g'(1) > 1$ then the smaller root q of Eq. (A1.4.4) is in the interval $0 \leq s < 1$, and if $E(\xi) = g'(1) \leq 1$ then $q = 1$ is the smaller root (see Fig. A1.4.1). We arrived at

THEOREM A1.4.2. *If the mathematical expectation $E(\xi)$ of the offspring number of an individual is ≤ 1 then the population dies out with probability 1; if it is ≥ 1 then the population survives with the positive probability $1 - q$.*

Appendix 2

ORDINARY DIFFERENTIAL EQUATIONS

A2.1 Basic Facts

Let the function $f : \mathbb{R} \times \mathbb{R}^n \mapsto \mathbb{R}^n, (t, x) \mapsto f(t, x)$ be continuous along with its derivative with respect to the second variable $x \in \mathbb{R}^n$: $f'_x(t, x)$. The elements of the matrix f'_x are the partial derivatives f'_{ix_k} , $(i, k = 1, 2, ..., n)$ of the coordinates of f. The equation

$$\dot{x} = f(t, x) \tag{A2.1.1}$$

where the overdot denotes differentiation with respect to the variable t (usually, the time) is an *n-dimensional system of first order ordinary differential equations (ODE)*. A *solution* of (A2.1.1) is a C^1 function $\varphi : I \mapsto \mathbb{R}^n$ ($I = (\alpha, \beta) \subset \mathbb{R}$ an open interval) that satisfies the equation : $\dot{\varphi}(t) \equiv f(t, \varphi(t))$ and is such that either $\beta = \infty$ ($\alpha = -\infty$) or if $\beta < \infty$ ($\alpha > -\infty$) then $|\varphi(t)| \to \infty$ as $t \to \beta$ ($t \to \alpha$). If it is prescribed that the solution shall assume the value $x^0 \in \mathbb{R}^n$ at the moment t_0 we say that the *initial condition*

$$x(t_0) = x^0 \tag{A2.1.2}$$

is given.

THEOREM A2.1.1. *Under the preceding conditions the initial value problem (A2.1.1)-(A2.1.2) has one and only one solution.*

For the proof of this and most of the other theorems and statements in this Section see, for example, Rouche and Mawhin (1973). The solution satisfying the initial condition (A2.1.2) is denoted by $\varphi(t, t_0, x^0)$, that is, $\varphi(t_0, t_0, x^0) = x^0$. The solution as a function of the initial conditions $\varphi : \mathbb{R} \times \mathbb{R} \times \mathbb{R}^n \mapsto \mathbb{R}^n$ belongs to the C^1 class and, as a consequence, depends continuously on the initial values.

The space $\mathbb{R}^n$ is called the *phase space* and $\mathbb{R} \times \mathbb{R}^n$ the *extended phase space* of the system. The graph of the solution φ: $\{(t, \varphi(t)) \in \mathbb{R} \times \mathbb{R}^n \mid t \in I\}$ is called the *integral curve* of the solution, its projection to the phase space: $\{x \in \mathbb{R}^n \mid x = \varphi(t), t \in I\}$ is the *path (trajectory)* of the solution. Theorem A2.1.1 means (geometrically) that through each point of the extended phase space there passes one and only one integral curve. If for an $a \in \mathbb{R}^n$ we have $f(t, a) \equiv 0$, then the constant solution $x \equiv a$ is called an *equilibrium (fixed point)* of the system. Its path is a point of the phase space.

If the function $f \in C^1$ does not depend on the time t then

$$\dot{x} = f(x) \tag{A2.1.3}$$

is called an *autonomous system*. It is easy to see that if φ is a solution of (A2.1.3) then for any $\delta \in \mathbb{R}$ the function $\psi(t) := \varphi(t + \delta)$ also is a solution and this solution has the same path in $\mathbb{R}^n$. We denote the solution of (A2.1.3) that assumes the initial value x^0 at $t = 0$ by $\varphi(t, x^0) := \varphi(t, 0, x^0)$. For any $t_0 \in \mathbb{R}$ the solution that assumes the same x^0 at t_0 is, clearly, $\varphi(t - t_0, x^0) \equiv \varphi(t, t_0, x^0)$. This shows that in the case of an autonomous system there is uniqueness not only in the extended phase space but also in the phase space in the sense that through each point of the phase space there passes one and only one path (the path of many solutions). This also means that for an autonomous system the initial time has no importance; at *any* time the system is in the state x^0 it will move along the same trajectory the same way.

The set $H \subset \mathbb{R}^n$ is said to be *invariant* with respect to system (A2.1.3) if $x^0 \in H$ implies that $\varphi(t, x^0) \in H$ for all t in the domain of φ. It is said to be *positively invariant* if $x^0 \in H$ implies $\varphi(t, x^0) \in H$ for $t > 0$. The set $\{x \in \mathbb{R}^n \mid x = \varphi(t, x^0), t \geq 0, t \in I\}$ is said to be the *positive semitrajectory* of this solution. A closed path is the trajectory of a periodic solution. We say that it is a *limit cycle* if it is the limit set of some trajectories as t tends to plus or minus infinity.

The C^1 function $V : \mathbb{R}^n \mapsto \mathbb{R}$ is said to be a *first integral* of Eq. (A2.1.3) if it is constant along every trajectory of the system, that is, if for any solution φ we have $V(\varphi(t)) \equiv$ constant. The following theorem can be proved easily.

THEOREM A2.1.2. *The C^1 function V is a first integral of system (A2.1.3) if and only if it satisfies the partial differential equation*

$$f(x) \cdot \text{grad}\, V(x) = 0 \, .$$

Let $A(t)$ be an $n \times n$ matrix whose entries are continuous functions for $t \in \mathbb{R}$ and consider the *homogeneous linear system of differential equations*

$$\dot{x} = A(t)x \ . \tag{A2.1.4}$$

This system has n linearly independent solutions $\varphi^1(t), \varphi^2(t), ..., \varphi^n(t)$ such that any solution φ of the system can be written uniquely in the form

$$\varphi(t) = \sum_{k=1}^{n} c_k \varphi^k(t)$$

where the c_k are real or complex constants. A system of n linearly independent solutions $\varphi^1(t), \varphi^2(t), ..., \varphi^n(t)$ is called a *basis* of (A2.1.4). Considering the solutions $\varphi^k(t)$ as column vectors we may form the $n \times n$ matrix $\Phi(t) := [\varphi^1(t), \varphi^2(t), ..., \varphi^n(t)]$. This is a regular matrix and it is a matrix solution of system (A2.1.4): $\dot{\Phi}(t) \equiv A(t)\Phi(t)$. Every solution φ of (A2.1.4) can be written in the form $\varphi(t) = \Phi(t)c$ with a suitably chosen constant column vector c. Such a regular matrix solution is called a *fundamental matrix* of the system.

If A is an $n \times n$ constant matrix then the *homogeneous linear system with constant coefficients*

$$\dot{x} = Ax \tag{A2.1.5}$$

can be solved explicitly. The *characteristic equation* of (A2.1.5) is $\det(A - \lambda I) = 0$ where I is the unit matrix. The roots $\lambda_1, \lambda_2, ..., \lambda_n$ of the characteristic equation are called the *eigenvalues* of the system. Denote by s^k an *eigenvector* corresponding to λ_k; then $s^k \exp(\lambda_k t)$ is a solution. If all the eigenvalues are simple then $s^1 \exp(\lambda_1 t), s^2 \exp(\lambda_2 t), ..., s^n \exp(\lambda_n t)$ is a basis of system (A2.1.5).

In many applications and especially in biology current behavior, the dynamics of the system, depends not only on the current state but also on the state of the system some time ago. Modeling such a situation leads us to *retarded functional differential equations (RFDE)* (see Hale 1977). We shall look only at the simplest case here. If the current (at time t) dynamics of the system is influenced also by its state at time $t - \tau$, $\tau > 0$ then the system has the form

$$\dot{x}(t) = f(t, x(t), x(t - \tau)) \ ,$$

where $f : \mathbb{R} \times \mathbb{R}^n \times \mathbb{R}^n \mapsto \mathbb{R}^n$ is continuous along with its derivatives with respect to its second and third variables. Here τ is called the *delay* or the *time lag*. In order to have a solution in some interval $(0, \beta)$, $\beta > 0$ we have to know the solution in $[-\tau, 0]$. This means that we have to attach an *initial function* $\phi \in C^0([-\tau, 0], \mathbb{R}^n)$ as an initial condition to the system. Under these conditions the system has a unique *solution* that is continuous on $[-\tau, \beta)$ for some $\beta > 0$, is equal to ϕ on $[-\tau, 0]$, and satisfies the system on $[0, \beta)$.

Consider the homogeneous linear system with delay

$$\dot{x}(t) = Ax(t) + Bx(t - \tau) \ , \tag{A2.1.6}$$

where A and B are $n \times n$ constant matrices. We may suppose that this system has a solution of the form $s \exp(\lambda t)$ where $s \neq 0$ is an n-dimensional constant column vector. Substituting this into (A2.1.6) gives the condition

$$\left(A + Be^{-\lambda\tau} - \lambda I\right) s = 0 \, .$$

This system has nontrivial solutions iff

$$\det\left(A + Be^{-\lambda\tau} - \lambda I\right) = 0 \, . \tag{A2.1.7}$$

This is called the *characteristic equation* of the delay system (A2.1.6). This is no longer a polynomial equation and, in general, has an infinite number of roots, that is, the system has an infinite number of *eigenvalues* (in other words its *spectrum* is an infinite set).

A2.2 Stability

In this Section we treat the stability concepts and conditions of autonomous systems of differential equations with emphasis on the stability of equilibria and periodic solutions. For the proofs and more details see Rouche and Mawhin (1973) and Rouche, et al. (1977).

Let the function $f : \mathbb{R}^n \mapsto \mathbb{R}^n$ be in the C^1 class and let $\psi : [0, \infty) \mapsto \mathbb{R}^n$ be a solution of the system

$$\dot{x} = f(x) \, . \tag{A2.2.1}$$

DEFINITION A2.2.1. We say that ψ is *stable in the Lyapunov sense* (or *neutrally stable)* if there is a $\rho > 0$ such that $\left|x^0 - \psi(0)\right| < \rho$ implies that the solution $\varphi(t, x^0)$ is defined on $[0, \infty)$, and for every $\varepsilon > 0$ there is a $\delta(\varepsilon) > 0$ such that $\left|x^0 - \psi(0)\right| < \delta(\varepsilon)$ implies $\left|\varphi(t, x^0) - \psi(t)\right| < \varepsilon$ for $t > 0$.

DEFINITION A2.2.2. We say that ψ is *attractive* if there is an $\eta > 0$ such that $\left|x^0 - \psi(0)\right| < \eta$ implies $\left|\varphi(t, x^0) - \psi(t)\right| \to 0$ as $t \to \infty$.

DEFINITION A2.2.3. We say that ψ is *asymptotically stable* if it is stable in the Lyapunov sense and it is attractive; we say that ψ is *globally asymptotically stable* if it is asymptotically stable and is attractive for all $x^0 \in \mathbb{R}^n$.

Because the "distance" of the solutions from the the fixed solution ψ plays a crucial role in these definitions, it makes sense to perform the transformation of the state variables $y = x - \psi(t)$ and ask what is the differential equation satisfied by y if the function x satisfies (A2.2.1)? We obtain

$$\begin{aligned} \dot{y}(t) &= \dot{x}(t) - \dot{\psi}(t) = f(\psi(t) + y(t)) - f(\psi(t) \\ &= f'(\psi(t))y(t) + o(|y(t)|) \, . \end{aligned}$$

Here f has been expanded into Taylor series and truncated at the first-order terms. If the higher-order terms are dropped for small variations $|y| = |x - \psi(t)|$ an approximate linear differential system is obtained for y :

$$\dot{y} = f'(\psi(t))y \,, \tag{A2.2.2}$$

which is called the *variational system with respect to the solution* ψ. Often one is able to determine the stability properties of the solution ψ of (A2.2.1) from the behavior of system (A2.2.2). This is the case in particular if $\psi(t) \equiv a$ is an equilibrium, that is, if $f(a) = 0$. In this case (A2.2.2) is a homogeneous linear system with constant coefficients, where the coefficient matrix is $f'(a) = [f'_{ix_k}(a)]$. The following theorem is referred to as the *theorem on stability by linearization.*

THEOREM A2.2.1. *The equilibrium $x = a$ of (A2.2.1) is asymptotically stable if the matrix $f'(a)$ is stable, that is, if the real parts of all its eigenvalues are negative; the equilibrium is unstable if the real part of at least one eigenvalue is positive.*

If the dimension of the system (A2.2.1) is two then $f'(a)$ is a 2×2 matrix and the following theorem is a simple consequence of Theorem A1.1.2:

THEOREM A2.2.2. *The two-dimensional matrix $f'(a)$ is stable iff* $\text{Trace} f'(a) < 0$ and $\det f'(a) > 0$.

For linear systems with constant coefficients we may say much more. Consider the system

$$\dot{x} = Ax \tag{A2.2.3}$$

where A is a constant $n \times n$ matrix.

THEOREM A2.2.3. *The equilibrium $x = 0$ of (A2.2.3) is asymptotically stable (and then globally) iff the matrix A is stable; if the real parts of the eigenvalues of A are nonpositive and those whose real part is zero are simple then $x = 0$ is stable in the Lyapunov sense.*

It is worthwhile to have a look at the phase portrait of (A2.2.3) in the most important generic cases when the dimension is the lowest relevant one, that is, $n = 2$. The system is written in the form

$$\begin{aligned} \dot{x}_1 &= ax_1 + bx_2 \\ \dot{x}_2 &= cx_1 + dx_2 \,. \end{aligned} \tag{A2.2.4}$$

If both eigenvalues of the coefficient matrix are real, different, and of the same sign, the equilibrium $x = 0$ is called a *node* (stable or unstable depending upon whether the common sign is negative or positive); if the two eigenvalues are real and of different sign then the origin is called a *saddle point;* if the eigenvalues are conjugate complex numbers with nonzero real part then the origin is a *spiral point* (stable or unstable depending upon whether the common real part is negative or positive); if the eigenvalues are conjugate pure imaginary numbers then the origin is a *center.* Figure A2.2.1 shows the respective phase portraits: (A) a stable node; (B) a saddle point; (C) a stable spiral point; (D) a center. A stable node and a stable spiral point are obviously asymptotically stable equilibria, a saddle point is unstable, and a center is stable in the Lyapunov sense (but not asymptotically).

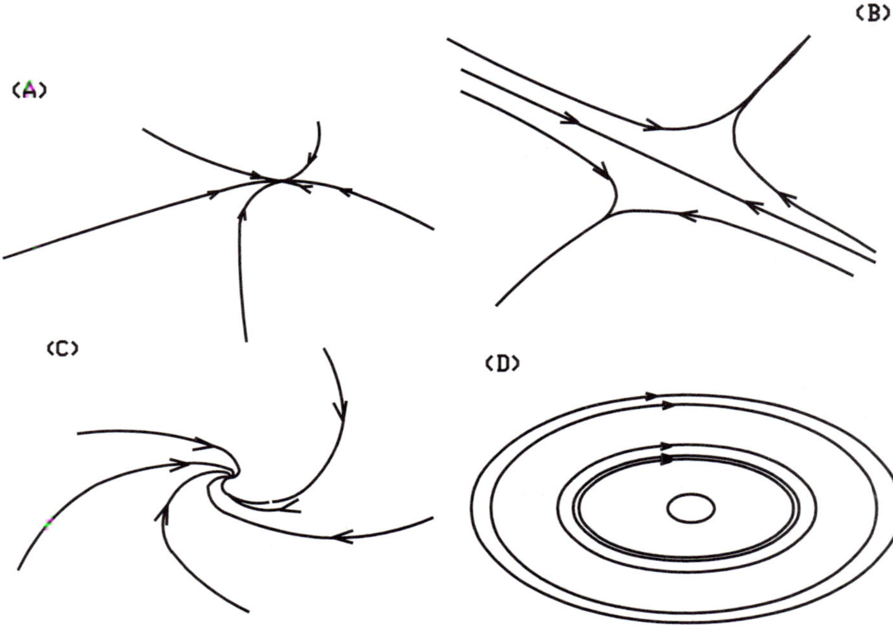

Figure A2.2.1: Phase portrait of system of equations (A2.2.4); (A) stable node, $a=-2$, $b=5$, $c=0.5$, $d=-2$; (B) saddle point, $a=-1$, $b=2$, $c=2$, $d=-1$; (C) stable spiral point, $a=-1$, $b=1$, $c=-1$, $d=-1$; and (D) center, $a=0$, $b=1$, $c=-1$, $d=0$ (PHASER).

We note that if system (A2.2.4) is the linearization of a 2D nonlinear system at an equilibrium $x=a$ and the origin is a node, a saddle or a spiral point, then the phase portrait of a nonlinear system is similar to that of (A2.2.4) *in a neighborhood* of the point a, and this equilibrium is also called a node, a saddle, or a spiral point, respectively. This is *not* the case with a center of the linearized system; in the latter case the equilibrium of the nonlinear system might be a center (meaning that in any neighborhood there is an infinite number of closed trajectories), or a stable or unstable spiral point.

For autonomous systems with delay, stability and asymptotic stability can be defined in an analogous way by applying a suitable norm for the difference of initial *functions.* If then (A2.1.6) is the linearization of a delay system at an equilibrium a theorem analogous to Theorem A2.2.1 can be proved: the negativity of the real parts of all the roots of the characteristic equation (A2.1.7) ensure the asymptotic stability of the equilibrium.

Let $x = a$ be an asymptotically stable equilibrium of system (A2.2.1). The open set $B \subset \mathbb{R}^n$ that contains a in its interior and is such that for $x^0 \in B$ we have $\varphi(t, x^0) \to a$ as t tend to infinity is called the *basin* or the *region of attractivity* of a. If the basin is the whole space then the equilibrium is globally attractive. The condition of Theorem A2.2.1 that ensures asymptotic stability does not tell us anything about the basin, although, it makes a big difference whether the basin is small or large. Another drawback of this theorem is that it does not settle the stability problem in the *critical case* where in addition to eigenvalues with a negative real part we have eigenvalues with a zero real part. The method that has the possibility of solving these problems is called *Lyapunov's direct method,* or the *method of Lyapunov functions.*

Let $D \subset \mathbb{R}^n$ be an open set containing the origin and $V : D \mapsto \mathbb{R}$ a C^1 function. We say that V is *positive definite* if $V(x) \geq 0$ for $x \in D$ and is zero only at $x = 0$; the function V is called *positive semidefinite* if $V(x) \geq 0$ for $x \in D$; we say that it is *indefinite* if it assumes positive and negative values as well. *Negative definity* and *semidefinity* are defined analogously. *The derivative of V with respect to system (A2.2.1)* at $x \in D$ is defined the following way–the solution $\varphi(t)$ of the system passing through point x is substituted into V and the composite function is differentiated with respect to t :

$$\begin{aligned} \dot{V}_{(A2.2.1)}(x) &= \frac{d}{dt}(V \circ \varphi)(t) = \operatorname{grad}V(\varphi(t)) \cdot \dot{\varphi}(t) \\ &= \operatorname{grad}V(x) \cdot f(x) = \sum\nolimits_{k=1}^{n} V'_{x_k}(x) f_k(x) \,. \end{aligned}$$

A function $V \in C^1(D)$ having some definiteness property along with its derivative with respect to the system is called a *Lyapunov function.*

We formulate the three fundamental theorems of Lyapunov's direct method. In the next four theorems it will be assumed without restriction on generality that the origin is an equilibrium of system (A2.2.1), that is, $f(0) = 0$. This can always be achieved by displacing the origin into the equilibrium point.

THEOREM A2.2.4. *If in a neighborhood of the origin a positive definite Lyapunov function can be found whose derivative with respect to the system is negative definite then the origin is asymptotically stable.*

THEOREM A2.2.5. *If in a neighborhood of the origin a positive definite Lyapunov function can be found whose derivative with respect to the system is negative semidefinite then the origin is stable in the Lyapunov sense.*

THEOREM A2.2.6. *If a Lyapunov function can be found whose derivative with respect to the system is negative definite, and the function itself assumes negative values in every neighborhood of the origin then the origin is unstable.*

Note that analogous, somewhat more sophisticated theorems hold for equilibria of nonautonomous systems. Theorem A2.2.4 allows for an estimate of the basin of the origin provided that the domain of the Lyapunov function is positively invariant. The following theorem due to Barbashin and Krasovskiy is often useful (see e.g. Farkas, 1994).

THEOREM A2.2.7. *If in a neighborhood of the origin a positive definite Lyapunov function $V : D \mapsto \mathbb{R}$ can be found whose derivative with respect to the system is negative semidefinite but the set $M := \{x \in D \mid \dot{V}_{(A2.2.1)}(x) = 0\}$ where the derivative is zero does not contain any positive semitrajectory apart from the origin then the origin is asymptotically stable.*

In the rest of this Section we treat the stability problem of a nonconstant periodic solution of the autonomous system (A2.2.1) in a concise way. Let $p(t) \neq$constant be a solution of (A2.2.1) and suppose that there exists a $T > 0$ such that $p(t+T) = p(t)$ for every t. Then p is a periodic solution with period T. Its path (or *orbit*) will be denoted by $\gamma := \{x \in \mathbb{R}^n \mid x = p(t), t \in \mathbb{R})$. If (A2.2.1) is a planar system, that is, the dimension $n = 2$, there is a comprehensive theory of periodic solutions, called the Poincaré-Bendixson Theory (see e.g., Farkas, 1994), with conditions for the existence of closed orbits. This theory implies, for instance, that inside a closed orbit of a planar system there must be at least one equilibrium point. It is easy to see that the concepts of Lyapunov or asymptotic stability are not suitable here, because, for example, for sufficiently small $\alpha > 0$ the difference of the initial values of the solutions $p(t+\alpha)$ and $p(t)$ can be made arbitrarily small and $p(t+\alpha) - p(t)$ still will not tend to zero. The applicable concepts are given in the following definition.

DEFINITION A2.2.4. We say that the periodic solution p of (A2.2.1) is *orbitally stable* if there is a $\rho > 0$ such that distance $(x^0, \gamma) < \rho$ implies that the solution $\varphi(t, x^0)$ is defined for all $t > 0$ and for every $\varepsilon > 0$ there exists a $\delta(\varepsilon) > 0$ such that distance$(x^0, \gamma) < \delta(\varepsilon)$ implies distance$(\varphi(t, x^0), \gamma) < \varepsilon$ for every $t > 0$; we say that p is *orbitally asymptotically stable* if it is orbitally stable and there exists an $\eta > 0$ such that distance$(x^0, \gamma) < \eta$ implies that distance$(\varphi(t, x^0), \gamma) \to 0$ as t tends to infinity.

In brief, these definitions mean that not the solution itself but its orbit is stable or asymptotically stable.

There is an important condition that guarantees orbital asymptotic stability by linearization but it needs some preparation. Consider the variational system with respect to the periodic solution p :

$$\dot{y} = f'(p(t))y \tag{A2.2.5}$$

and its fundamental matrix $\Phi(t)$, which assumes the unit matrix at $t = 0 : \Phi(0) = I$. The value of this matrix at $t = T$ is called the *principal matrix* of the variational Eq. (A2.2.5) and the eigenvalues of the principal matrix $\Phi(T)$ are called the *characteristic multipliers.* It is fairly easy to see that the number 1 is one of the characteristic multipliers. The following theorem is due to Andronov and Witt (for the proof see, e.g., Farkas, 1994).

THEOREM A2.2.8 *If $n - 1$ characteristic multipliers of system (A2.2.5) are in modulus < 1 then the periodic solution p of system (A2.2.1) is orbitally asymptotically stable.*

Beyond equilibria and periodic solutions the stability concepts can be extended to arbitrary compact sets. If the compact invariant set $H \subset \mathbb{R}^n$ has a neighborhood such that for every initial value in this neighborhood the corresponding solution tends to H as t tends to infinity (minus infinity) then H is called an *attractor* (*repellor*) of the system.

A2.3 Bifurcations

When modeling dynamical phenomena in biology by differential equations most circumstances are considered to have relatively little effect on the dynamics and are neglected. If the model is an n-dimensional autonomous system of differential equations

$$\dot{x} = f(x) \tag{A2.3.1}$$

where $f \in C^1$, which generates the *flow* $\varphi(t, x)$ taking the system from the state x at time 0 into $\varphi(t, x)$ at time t (here, as before, φ is the solution with initial value x), then one can be sure that the "real system" to be solved and the "real flow" is different from what we have. One hopes that the difference is small and the qualitative behavior of the solutions, that is, of the flow belonging to Eq. (A2.3.1) does not change if f is replaced by a vector field "near to it." The related problem is the problem of *structural stability*, which is different from the stability problem with respect to the change of the initial values as treated in the previous Section. When we say that the solutions φ as functions of the initial values represent a flow we assume that all solutions are defined on

$(-\infty, \infty)$. This is not a restriction on generality because if system (A2.3.1) is replaced by

$$\dot{x} = f(x)/(1 + f^2(x))^{1/2},$$

the last one has this property (the modulus of the right-hand side is everywhere < 1) and it has the same trajectories traversed in the same direction.

DEFINITION A2.3.1. We say that two flows are *equivalent* if there is a homeomorphism (a continuous one-to-one mapping of the space onto itself whose inverse is also continuous) that carries the trajectories of one system into those of the other and preserves the direction of the motion along them.

DEFINITION A2.3.2. We say that the flow generated by system (A2.3.1) is *structurally stable* if it is equivalent to every flow "near to it."

The *distance of the flows* is defined as the distance of the vector fields on the right-hand sides of the respective generating differential equations; the *distance* of two C^1 vector fields is the modulus of their difference plus the norm of the difference of their derivatives. Unfortunately, this concept of distance works only if the flow is acting on a compact manifold and not on $\mathbb{R}^n$ (on $\mathbb{R}^n$ not every two flows have a distance). Although the situation can be saved if we restrict the flows to compact subsets of $\mathbb{R}^n$, we have to suppose that the flow is transversal to the boundary of the set; this means that every path crosses the boundary transversally (e.g., from the outside to the inside).

Two basic problems are to be attacked–first, how can structurally stable flows be characterized and second, whether are they generic in the space of C^1 systems? If the dimension n of the system is *two* then the characterization problem has already been solved by Andronov and Pontryagin (see e.g. Farkas, 1994) and later by Peixoto (1959, 1962). It turned out that among other things the loss of structural stability is related to the occurrence of *heteroclinic trajectories* joining saddle points. We say that a path is *heteroclinic* if it "joins two equilibria" in the sense that it tends to the first equilibrium as t tends to minus infinity and tends to the second one as t tends to plus infinity. If the two equilibria are the same then the path is said to be *homoclinic*. Peixoto also proved that structurally stable systems are *generic* in dimension two, that is, they form an open and dense set in the space of C^1 systems. If the dimension is higher than two then there is no general characterization and we know that structurally stable systems are not generic.

The situation is different in the special case of *gradient systems*. (A2.3.1) is said to be a gradient system if f is the (negative) gradient of a scalar function $V \in C^2(\mathbb{R}^n)$: $f(x) = -\text{grad}V(x)$. It is easy to prove, for instance, that a gradient system can have neither a closed path (a nonconstant periodic solution) nor a homoclinic trajectory. Smale (1961) characterized structurally stable gradient systems and proved that structurally stable systems are generic in the space of gradient systems.

The problem of resistance of the qualitative properties of a flow to small changes to the right-hand side of the system is easier to handle if instead of looking at the whole space of flows we restrict the attention to members of a family of systems depending on some parameters. Here we treat only the case of a single parameter. Consider the family of systems

$$\dot{x} = f(x, \mu) \ , \ x \in \mathbb{R}^n \ , \ \mu \in \mathbb{R} \ , \ f \in C^1 \ . \tag{A2.3.2}$$

At a certain value of the parameter μ the flow generated by the system has some qualitative properties including the number of equilibria, their stability, the number of limit cycles, etc. We expect that for small changes of the parameter these properties persist and indeed this is usually the case. However, there might be values of the parameter where this is not the case–arbitrarily small changes in parameter results in a change in properties, the number of equilibria or limit cycles changes, a formerly stable one becomes unstable, etc. If this happens we say that the system has undergone a *bifurcation* at that parameter value and the latter is a *bifurcation point.* For a detailed study of the theory of bifurcations see, for example, Guckenheimer and Holmes (1983) and Chow and Hale (1982). Here we present, first, two generic bifurcations involving only equilibria that occur in systems of a single bifurcation parameter and then the generic Andronov-Hopf bifurcation, which gives rise to limit cycles.

The saddle-node bifurcation. Consider the one parameter family of 1D systems

$$\dot{x} = -x^2 + \mu \ . \tag{A2.3.3}$$

For $\mu < 0$ this equation has no equilibrium. At $\mu = 0$ the equilibrium $x = 0$ appears; it is called a *saddle-node* because it attracts solutions with positive initial values and repels those with negative ones (if we attached the equation $\dot{y} = -y$ to a 2D system then the node and the saddle property, respectively, would be apparent from one side and then the other). For $\mu > 0$ the equation has two equilibria: $x_1 = \sqrt{\mu}$ and $x_2 = -\sqrt{\mu}$; the first one is asymptotically stable and the second one is unstable. See Fig. A2.3.1.

The situation in this simple example is generic; if a system has no equilibrium, a parameter is varied, and at a value of the parameter equilibria appear then they do this, usually, in pairs, with one stable and the other unstable.

The pitchfork bifurcation. Consider the one parameter family of 1D systems

$$\dot{x} = -x^3 + \mu x \ . \tag{A2.3.4}$$

The right-hand side is an odd function, so that $x = 0$ is always an equilibrium. At $x = 0$ the derivative of the right-hand side $(-x^3 + \mu x)' = -3x^2 + \mu$ is negative for $\mu < 0$ and positive for $\mu > 0$. As a consequence, this equilibrium is asymptotically stable for $\mu < 0$ and it is unstable for $\mu > 0$. For $\mu = 0$ it is still asymptotically stable, as this can be seen by integrating the equation (but

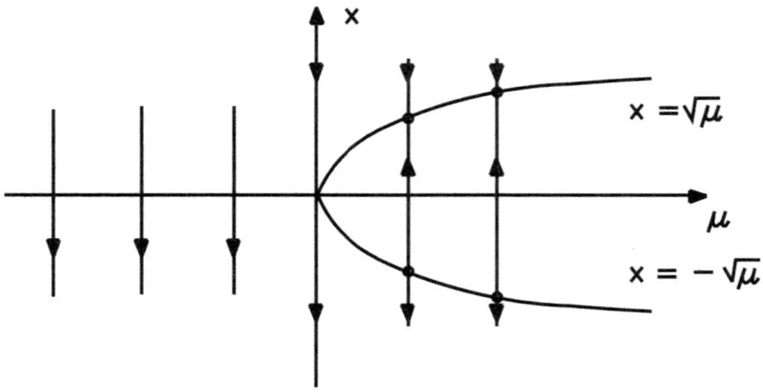

Figure A2.3.1: The saddle-node bifurcation in Eq. (A2.3.3).

it is no longer linearly stable). Thus, if the bifurcation parameter μ is increased at $\mu = 0$ this equilibrium loses its stability. At the same time the equation has two new equilibria for $\mu > 0$: $x_1 = \sqrt{\mu}$ and $x_2 = -\sqrt{\mu}$. If these values are substituted for x into the derivative of the right-hand side we obtain $-2\mu < 0$. Therefore, both new equilibria are asymptotically stable. See Fig. A2.3.2 (this is called a *supercritical* pitchfork bifurcation because the new stable equilibria appear above the critical bifurcation point $\mu = 0$).

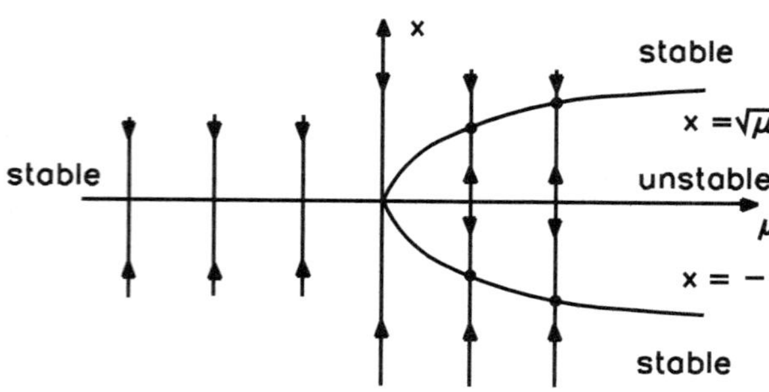

Figure A2.3.2: Supercritical pitchfork bifurcation in equation (A2.3.4).

The Andronov-Hopf bifurcation. For the pitchfork bifurcation, as the bifurcation parameter was increased through the value zero the eigenvalue of the system linearized at the origin turned from negative to positive, thereby causing destabilization of the origin. There is another generic way to destabilize an equilibrium for a system of at least two dimensions–this is when a pair of conjugate complex eigenvalues cross the imaginary axis from the left to the right half plane away from the origin. Such a loss of stability usually gives rise to a limit cycle of the system. The conditions under which this happens are expressed in the following theorem due to Andronov (1937) in dimension two and to Hopf (1942) in the n-dimensional case.

THEOREM A2.3.1. *Suppose that in system (A2.3.2)* $f \in C^5$, $f(0,\mu) \equiv 0$, *the matrix* $f'_x(0,\mu)$ *has a pair of conjugate complex eigenvalues* $\alpha(\mu) \pm i\omega(\mu)$, $\omega(\mu) > 0$, $\alpha(0) = 0$, *the derivative of the real part with respect to the parameter at zero is positive:* $\alpha'(0) > 0$ *and the other n-2 eigenvalues have negative real parts; then*

(i) there exists a $\delta > 0$ *and a function* $\mu : (-\delta,\delta) \mapsto \mathbb{R}$ *such that* $\mu \in C^3$, *and for* $\varepsilon \in (-\delta,\delta)$ *the system* $\dot{x} = f(x,\mu(\varepsilon))$ *has a periodic solution* $p(t,\varepsilon)$ *with period* $T(\varepsilon) > 0$, *also* $T \in C^3$, $\mu(0) = 0$, $T(0) = 2\pi/\omega(0)$, $p(t,0) \equiv 0$, *and the amplitude of this periodic solution (the average distance of its orbit from the origin) is proportional to* $\sqrt{|\mu(\varepsilon)|}$; *and*

(ii) the origin $(x,\mu) = (0,0)$ *of the space* $\mathbb{R}^n \times \mathbb{R}$ *has a neighborhood* $U \subset \mathbb{R}^n \times \mathbb{R}$ *that does not contain any periodic orbit of Eq. (A2.3.2) but those of the family* $p(t,\varepsilon)$, $\varepsilon \in (-\delta,\delta)$.

Thus, this theorem guarantees the existence of a closed (periodic) orbit for small $|\mu|$ and also the uniqueness provided that the function μ is strictly monotonous in $[0,\delta]$. In the generic case function $\mu(\varepsilon)$ is either positive or negative for every $\varepsilon \in (-\delta,\delta)$, $\varepsilon \neq 0$. In the first case the bifurcation is said to be *supercritical* and in the second *subcritical* (see Fig. A2.3.3).

In the supercritical case the periodic solutions appear when the equilibrium has lost its stability and they are orbitally asymptotically stable; in the subcritical case the periodic solutions are unstable and exist for μs when the equilibrium is still asymptotically stable. To determine whether the bifurcation is super- or subcritical is not an easy task (for the proof of the theorem and other details see, e.g., Farkas, 1994). If the right-hand side f of system (A2.3.2) is *analytic* the following criterion is true: *the bifurcation is supercritical iff for* $\mu = 0$ *the equilibrium* $x = 0$ *is asymptotically stable.*

An important property of equilibria to be taken into account when dealing with their bifurcations is the existence of certain "local invariant manifolds" attached to these points. We say that a manifold passing through an equilibrium point of system (A2.3.1) is a *local invariant manifold* if given an initial value on this manifold the path of the corresponding solution stays in this manifold as long as it remains in a certain neighborhood of the equilibrium point. Suppose that point a is an equilibrium of the system, that is, $f(a) = 0$ and that linearizing the system at this point the matrix $f'(a)$ has s, u and c eigenvalues with negative, positive and zero real parts, respectively ($s + u + c = n$). Then point a has an $s-$, $u-$ and $c-$dimensional local invariant *stable, unstable* or *center manifold, respectively:* W^s, W^u and M, respectively, such that if $x^s \in W^s$ or $x^u \in W^u$, respectively, then the solution $\varphi(t,x^s) \to a$, $t \to \infty$, or $\varphi(t,x^u) \to a$, $t \to -\infty$, respectively, and if a solution is such that $\varphi(t,x^c)$ stays in a neighborhood of a for $t \in (-\infty,\infty)$ then $x^c \in M$ (see Hartman, 1964). The local stable and unstable manifolds are uniquely determined but point a may have several center

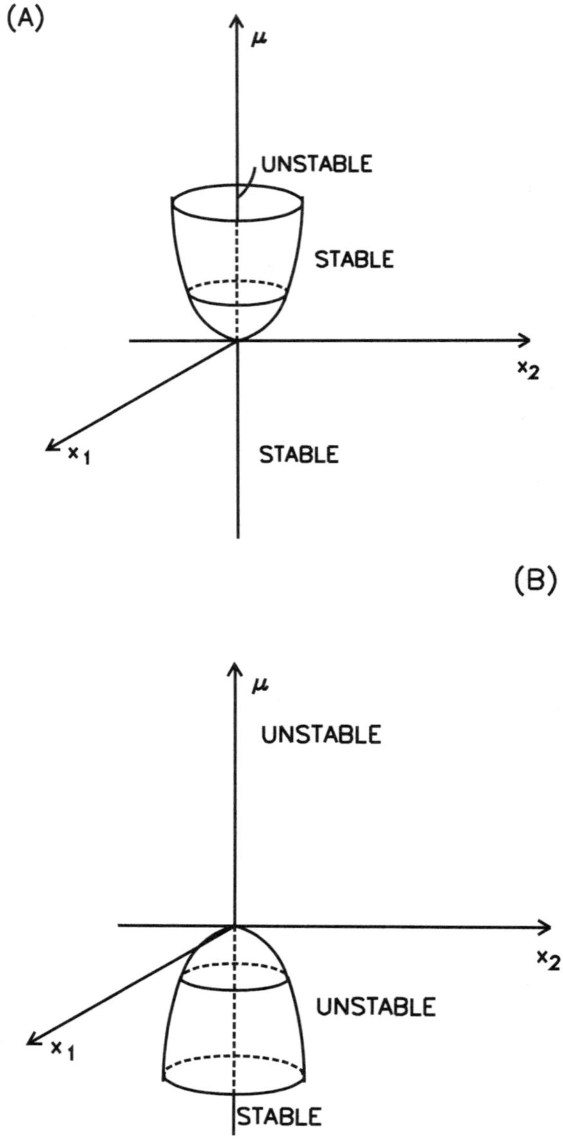

Figure A2.3.3: Supercritical (A) and subcritical (B) Andronov-Hopf bifurcation at $\mu = 0$.

manifolds with the same tangent space at a. Obviously, the center manifold contains the local closed orbits of the system if there are any.

Appendix 3

PARTIAL DIFFERENTIAL EQUATIONS

A3.1 First-Order Linear Equations

We present here the most important facts about first-order linear partial differential equations (PDE). As we are applying in this book these concepts and theorems for equations containing functions of only two variables, we shall restrict ourselves to this case (and to the case of three independent variables what is needed in the treatment).

The equation

$$f(x,y)\frac{\partial u}{\partial x}+g(x,y)\frac{\partial u}{\partial y}=0 \qquad \text{(A3.1.1)}$$

is called a *first-order homogeneous linear partial differential equation.* Here it is assumed that f and g are smooth functions, that is, of the class C^1 in an open domain $D \subset \mathbb{R}^2$ and that $f^2+g^2>0,\ (x,y)\in D$. The task is to find a C^1 function $u(x,y)$ defined in D or on a subset of D that satisfies the PDE. The graph of a solution u of Eq. (A3.1.1) is a surface in the 3D space whose equation (in the orthogonal Cartesian coordinate system x,y,z) is $z=u(x,y)$. Such a surface is called an *integral surface* of the equation. *"Boundary conditions"* usually are attached to the equation, meaning that the values of the solution are prescribed along a curve in D or, rather, on the boundary of D. For the sake of simplicity we are going to assume that the axis x is the boundary or part of the boundary of D and the conditions are given along it or along an interval on it, $u(x,0)=U(x)$ where U is a given smooth function. Geometrically this means that a curve is given above the axis x and we are looking for a surface that contains this curve and is the graph of a solution at the same time. Now, if c is a constant in the range of the solution u then $u(x, y) = c$ is the equation of a *level curve* of u, and grad $u = [u'_x, u'_y]$ is orthogonal to this level curve at each point. With Eq. (A3.1.1) this means that the vector $[f, g]$ is the tangent vector of the level curve. In some smooth parametrization the equation of this level curve can be given by $x = \varphi(t)$, $y = \psi(t)$ where these functions satisfy the system of ordinary differential equations

$$\dot{x} = f(x, y) , \qquad \dot{y} = g(x, y) . \tag{A3.1.2}$$

This system is called the *characteristic system* of (A3.1.1). The curves $x = \varphi(t)$, $y = \psi(t)$, $z = c$ are the *characteristics* of (A3.1.1); as c varies along the range of u these curves build up the integral surface. Their projections to the x, y plane, that is, the level curves of u are the *projected characteristics*.

The following statements can be proved easily.

A constant function is always a solution of Eq. (A3.1.1).

If an integral surface has a common point with a characteristic then it contains this characteristic.

If two integral surfaces have a common point then they intersect in the characteristic passing through this point.

A smooth function u is a solution of Eq. (A3.1.1) iff it is constant along each projected characteristic.

If $u(x, y)$ is a solution and Φ is a smooth function defined on the range of u then the composite function $\Phi \circ u$ is a solution too.

A smooth function is a solution of Eq. (A3.1.1) iff it is a first integral of the system (A3.1.2) (see Appendix 2.1).

For the local existence and uniqueness of solutions the following theorem holds.

THEOREM A3.1.1. *If a smooth curve γ transversal to the projected characteristics is given in D and a smooth real valued function is defined along it then in a neighborhood of γ Eq. (A3.1.1) has one and only one solution that assumes the prescribed values along γ.*

The condition that γ is transversal to the characteristics means that at no point is its tangent parallel to the characteristic direction $[f, g]$. No proof of this theorem will be given here (for a proof see, e.g., Evans, 1998) but the method of solution of such a boundary value problem will be described. Suppose curve γ is an interval I on the axis x, the vector $[f, g]$ is not parallel to the axis x at points of I, that is, $g(x, 0) \neq 0$, $x \in I$, and $U : I \mapsto \mathbb{R}$ is a smooth function. For each $x^0 \in I$ the characteristic system (A3.1.2) is solved as $(\varphi(t, x^0, 0), \psi(t, x^0, 0))$, which is the solution that satisfies $(\varphi(0, x^0, 0), \psi(0, x^0, 0)) = (x^0, 0)$ and is defined for $t \in [0, T(x^0))$ with some $T(x^0) > 0$. The constant $U(x^0)$ is attached and this way we obtain the characteristic passing through the point $(x^0, 0, U(x^0))$; its parametric equation is $(x, y, z) = (\varphi(t, x^0, 0), \psi(t, x^0, 0), U(x^0))$. Since the mapping $(t, x^0) \mapsto (\varphi(t, x^0, 0), \psi(t, x^0, 0))$ is one-to-one (because trajectories of autonomous systems do not intersect) from $(x, y) = (\varphi(t, x^0, 0), \psi(t, x^0, 0))$, in principle we may express t and x^0 uniquely as functions of x and y. Substituting the function $x^0(x, y)$ obtained this way into U makes the function $u(x, y) := U(x^0(x, y))$ the required solution of Eq. (A3.1.1).

EXAMPLE. Consider the equation

$$y\frac{\partial u}{\partial x} - x\frac{\partial u}{\partial y} = 0$$

with the boundary condition $u(x, 0) = U(x) = x^2$ along axis x. Here $f(x, y) = y$, $g(x, y) = -x \neq 0$ if $x \neq 0$. The characteristic system is $\dot{x} = y$, $\dot{y} = -x$. The solution with initial values $(x^0, 0)$ is obviously $(x^0 \cos t, -x^0 \sin t)$. From $x = x^0 \cos t$, $y = -x^0 \sin t$ we obtain $x^0 = \sqrt{x^2 + y^2}$. By substituting the last expression into the boundary condition, the required solution is $u(x, y) = U\left(\sqrt{x^2 + y^2}\right) = x^2 + y^2$. The corresponding integral surface is the rotational paraboloid $z = x^2 + y^2$.

We shall treat the case of three independent variables by analogy. Consider the homogeneous linear first-order PDE

$$f(x, y, z)\frac{\partial v}{\partial x} + g(x, y, z)\frac{\partial v}{\partial y} + h(x, y, z)\frac{\partial v}{\partial z} = 0 \qquad \text{(A3.1.3)}$$

where f, g, h are smooth functions defined on an open domain $D \subset \mathbb{R}^3$. The *characteristic system* corresponding to Eq. (A3.1.3) is

$$\dot{x} = f(x, y, z)\ , \ \dot{y} = g(x, y, z)\ , \ \dot{z} = h(x, y, z)\ . \qquad \text{(A3.1.4)}$$

The solutions of this system are the *projected characteristics*. The smooth function v is a solution of Eq. (A3.1.3) iff it is a first integral of system (A3.1.4), that is, iff it is constant along the projected characteristics. If $(x, y, z) = (\varphi(t), \psi(t), \chi(t))$ is a solution of (A3.1.4) and c is a real constant then the curve $(\varphi(t), \psi(t), \chi(t), c)$ of $\mathbb{R}^4$ is a *characteristic* of Eq. (A3.1.3). The graph of a solution v is the union of such characteristics.

We may prescribe the values of a solution along a smooth surface of $\mathbb{R}^3$, which is transversal to the projected characteristics. Suppose that the plane $z = 0$ or part of it is contained in D, no characteristic direction is contained in this plane, that is, $h \neq 0$, and the boundary condition $v(x, y, 0) = V(x, y)$ is attached to Eq. (A3.1.3) where V is a smooth function. Then locally there exists a unique solution satisfying this condition and it can be determined the following way. For each admissible initial value $(x^0, y^0, 0)$ we determine the solution of system (A3.1.4):

$$(x, y, z) = (\varphi(t, x^0, y^0, 0), \psi(t, x^0, y^0, 0), \chi(t, x^0, y^0, 0))$$

and express t, x^0, y^0 as a function of x, y, z and substitute the functions $x^0(x,y,z)$, $y^0(x,y,z)$ into V. The function $V(x^0(x,y,z),\ y^0(x,y,z))$ will be the required solution.

The equation

$$f(x,y,z)\frac{\partial z}{\partial x} + g(x,y,z)\frac{\partial z}{\partial y} = h(x,y,z) \tag{A3.1.5}$$

where f, g, h are smooth functions in an open domain $D \subset \mathbb{R}^3$ is called a *first-order quasilinear partial differential equation* (in two independent variables). A function $u(x,y)$ is a solution if it is in the C^1 class, its graph $(x, y, u(x,y))$ is in D if (x,y) is in its domain, and if being substituted for z in Eq. (A3.1.5) makes it an identity.

The equation

$$f(x,y,z)\frac{\partial v}{\partial x} + g(x,y,z)\frac{\partial v}{\partial y} + h(x,y,z)\frac{\partial v}{\partial z} = 0 \tag{A3.1.6}$$

is called the *homogeneous equation corresponding to Eq. (A3.1.5).*

The solution of Eq. (A3.1.5) is reduced to the solution of the corresponding homogeneous equation according to the following

THEOREM A3.1.2. *Suppose that $v(x,y,z)$ is a solution of Eq. (A3.1.6), the function $u(x,y)$ is in the C^1 class, its graph is in D, the derivative $v'_z(x,y,u(x,y)) \neq 0$ apart from a set of measure zero, and $v(x,y,u(x,y)) \equiv c$ (a constant); then $u(x,y)$ is a solution of Eq. (A3.1.5).*

PROOF. Differentiating the identity $v(x,y,u(x,y)) \equiv c$ by x and y we obtain

$$\frac{\partial v}{\partial x} + \frac{\partial v}{\partial z}\frac{\partial u}{\partial x} = 0\,, \quad \frac{\partial v}{\partial y} + \frac{\partial v}{\partial z}\frac{\partial u}{\partial y} = 0$$

or

$$\frac{\partial v}{\partial x} = -\frac{\partial v}{\partial z}\frac{\partial u}{\partial x}\,, \quad \frac{\partial v}{\partial y} = -\frac{\partial v}{\partial z}\frac{\partial u}{\partial y}\,.$$

The substitution of the last expressions into Eq. (A3.1.6) yields

$$v'_z(x,y,u(x,y))\,(f(x,y,u(x,y))u'_x(x,y)$$

$$+g(x,y,u(x,y))u'_y(x,y) - h(x,y,u(x,y))) = 0\,.$$

As the first factor is almost everywhere nonzero, the second factor must be identically zero, and this proves the Theorem. □

Thus we see that if a solution $v(x, y, z)$ of Eq. (A3.1.6) is known and the Implicit Function theorem can be applied to $v(x, y, z) = c$, that is, z can be expressed as a uniquely determined smooth function, then this function is a solution of Eq. (A3.1.5). As we have already seen, the solution of Eq. (A3.1.6) is based on the solution of the characteristic system (A3.1.4). It follows now that a smooth function $u(x, y)$ is a solution of Eq. (A3.1.5) iff the corresponding integral surface $z = u(x, y)$ has the property that through each point of it there passes at least one projected characteristic of (A3.1.4) and if a projected characteristic has a point common with this surface then it is contained in it.

A3.2 Reaction-Diffusion Equations

Suppose that a spatial domain (in the 3D physical space, or in a membrane considered to be 2D, or in a thin tube considered to be 1D) is filled up with some liquid or gas that contains a certain substance (also a liquid or a gas, respectively), where the density of the latter at time t and at the point given in Cartesian orthogonal coordinates by (x, y, z) is $u(t, x, y, z)$. Function u is supposed to be nonnegative and of the C^2 class. The molecules (particles) of the substance perform random motions that become more vivid when the temperature is higher. As a consequence, the substance is moving, "diffusing" away from places where its density is high and towards places where its density is low–density is attempting to become uniform. Thus the substance flows towards the direction in which the decrease in density is more rapid than in other directions and is proportional to the highest rate of density diminution. This is *Fick's law.* According to this law, if the flow vector is denoted by $q(t, x, y, z) = [q_1(t, x, y, z), q_2(t, x, y, z), q_3(t, x, y, z)]$ then

$$q(t, x, y, z) = -d \operatorname{grad} u(t, x, y, z) = -d\,[u'_x, u'_y, u'_z], \tag{A3.2.1}$$

where d is the *diffusion coefficient.* The latter could be a function of time and place, as well as of the density itself or of the temperature but for the purposes of this book we consider it to be a positive constant. The physical meaning of the vector q is that the quantity of the substance passing through a small surface of unit surface area orthogonal to q in unit time is $|q|$. Let us choose an arbitrary spatial domain V in that part of the space that is filled up and suppose that the boundary of V is a smooth closed surface S with an outside directed normal vector. The total quantity of substance in V is

$$\iiint_V u(t, x, y, z)\,dx\,dy\,dz$$

and the increase of this quantity in unit time is the derivative of this integral with respect to time t. If we assume that the substance is neither produced nor decays in V, then this increase must be equal to the quantity flowing into V through S in unit time. This way we arrive at the balance equation:

$$\frac{d}{dt}\iiint_V u(t, x, y, z)\,dx\,dy\,dz = -\iint_S q(t, x, y, z)\,dS\,.$$

By rearranging, differentiating under the integral sign, substituting from Eq. (A3.2.1), and applying the Divergence theorem we obtain

$$\iiint_V \left(\frac{\partial u}{\partial t} - d\,\mathrm{div}\ \mathrm{grad}\, u\right) dxdydz = 0\ , \qquad \text{(A3.2.2)}$$

where div grad $u = \Delta u = \frac{\partial^2 u}{\partial x^2} + \frac{\partial^2 u}{\partial y^2} + \frac{\partial^2 u}{\partial z^2}$. As Eq. (A3.2.2) holds for arbitrary domains V the integrand must be zero:

$$\frac{\partial u}{\partial t} = d\,\Delta u\ . \qquad \text{(A3.2.3)}$$

This is the classical linear *partial differential equation of diffusion.* It is a *parabolic equation*; if proper boundary and initial conditions are attached then it has a uniquely determined solution defined on the spatial domain for positive values of the time. Assume for the sake of simplicity that the space is 1D and the diffusion takes place in the interval $[0, L]$ with some positive L. Then the equation takes the form

$$\frac{\partial u}{\partial t} = d\,\frac{\partial^2 u}{\partial x^2}\ . \qquad \text{(A3.2.3}_1\text{)}$$

On the boundary, that is, at the points $x = 0$, $x = L$ one may prescribe, say, *Neumann* (or *no flux) boundary conditions*, which express the fact that no substance is flowing into or out of the domain:

$$\frac{\partial u(t,0)}{\partial x} = \frac{\partial u(t,L)}{\partial x} = 0\ , \qquad t \geq 0\ . \qquad \text{(A3.2.4)}$$

For *initial conditions* we may prescribe the values of u at the start, that is, at $t = 0$:

$$u(0,x) = U(x)\ , \qquad x \in [0,L]\ , \qquad \text{(A3.2.5)}$$

where U is a known function of the C^2 class.

The classical method of solving the problem (A3.2.3$_1$)-(A3.2.4)-(A3.2.5) is called *Fourier's method,* or *the method of the separation of variables.* According to this we first solve problem (A3.2.3$_1$)-(A3.2.4), in which the equations are homogeneously linear by assuming that it has a solution of the form $u(t,x) = T(t)X(x)$. Substituting this into Eq. (A3.2.3$_1$) and dividing by dTX gives

$$\frac{\dot{T}(t)}{d\,T(t)} = \frac{X''(x)}{X(x)} = -\lambda\ ,$$

where the overdot and the double prime denote differentiation with respect to t and x, respectively, and λ must be a constant, because the first term depends only on t and the second only on x. Equation $X'' = -\lambda X$ is first solved. Solutions have to satisfy $X'(0) = X'(L) = 0$. These are clearly $X(x) = \cos(x\sqrt{\lambda})$, provided that $L\sqrt{\lambda} = k\pi$, $(k = 0, 1, 2, ...)$. Values $\lambda_k = (k\pi/L)^2$

are called the *eigenvalues* of the problem with corresponding *eigenfunctions* $X_k(x) = \cos(k\pi x/L)$. A differential equation for T belongs to each eigenvalue : $\dot{T} = -\lambda_k dT$. Its solution is $T_k(t) = \exp(-\lambda_k dt)$. As each function $u_k(t,x) = T_k(t)X_k(x)$ is a solution of the problem (A3.2.3$_1$)-(A3.2.4), their arbitrary linear combination also is one and such is a series of these functions provided that it can be substituted into Eq. (A3.2.3$_1$) :

$$u(t,x) = \sum_{k=0}^{\infty} c_k \exp(-(k\pi/L)^2 dt) \cos(k\pi x/L) \ . \tag{A3.2.6}$$

The coefficients of this series are to be determined from the initial condition (A3.2.5). One has to expand the function U into cosine Fourier series:

$$U(x) = \sum_{k=0}^{\infty} c_k \cos(k\pi x/L) \ ,$$

where

$$c_k = (2/L) \int_0^L U(x) \cos(k\pi x/L) dx \ , \ (k = 1, 2, 3, ...)$$

and

$$c_0 = (1/L) \int_0^L U(x) dx.$$

These values are to be substituted into Eq. (A3.2.6). It is clear that no matter which initial distribution U we have, all the terms in the series except the zeroth one tend to zero as t tends to infinity, so that, $u(t,x) \to c_0$, which is the integral mean of the initial distribution.

We note that one may proceed on in a similar way if other homogeneous linear boundary conditions are prescribed–for example, if by introducing the substance into the interval in a constant rate at the boundary, the density $u(t,0) = u(t,L)$ is kept constant there. In this case we speak about *Dirichlet boundary conditions.*

The situation is different if the substance has "sources" or "sinks" in the domain. If it is assumed that these are homogeneously distributed in the bounded, connected spatial domain $B \subset \mathbb{R}^3$ with smooth boundary S and produce or absorb the substance at a rate that does not depend on time and space but on the density itself, then Eq. (A3.2.3) is replaced by

$$\frac{\partial u}{\partial t} = d \, \Delta u + f(u) \ , \tag{A3.2.7}$$

where the "reaction" or the source term $f : \mathbb{R} \mapsto \mathbb{R}$ is of class C^1 and Δu is given as following Eq. (A3.2.2). This is called a *reaction-diffusion equation.* By dropping the diffusion term the ODE

$$\frac{du}{dt} = f(u)$$

is obtained (called the *kinetic equation* corresponding to Eq. (A3.2.7)). If the last equation has an equilibrium, $f(a) = 0$ then $u(t, x, y, z) \equiv a$ is also a solution of the PDE (A3.2.7). If $u = a$ is an asymptotically stable equilibrium of the kinetic equation and B is convex then, under no flux boundary conditions, it is also an asymptotically stable solution of Eq. (A3.2.7), which means that solutions with initial values near to a stay near to a and tend to a as t tends to infinity. In this case, when the underlying space is 3D, no flux boundary conditions mean that the directional derivative of u in the direction of the normal vector of S is zero everywhere on S for $t \geq 0$. Eq. (A3.2.7) also may have nonconstant stationary solutions, that is, solutions $u(x, y, z)$ that do not depend on t. However, under the forementioned conditions no such solution can be asymptotically stable. For these results see Casten and Holland (1977,1978). If there is more than one substance diffusing and reacting with another then the last statement is no longer valid. We shall handle this problem in the next Section.

A3.3 Turing Bifurcation

Consider now two substances that react with each other (activating or inhibiting the production of each other) and diffuse in a spatial domain according to Fick's law. We assume a 2D bounded, connected spatial domain B with piecewise smooth boundary ∂B, and denote the respective densities at time t and point $(x, y) \in B$ by $u(t, x, y)$, $v(t, x, y)$ where x and y are Cartesian orthogonal coordinates. Then proceeding analogously to the single substance case in the previous Section, the dynamics is described by the system of reaction-diffusion equations

$$\begin{aligned} \frac{\partial u}{\partial t} &= d_u \Delta u + f(u, v) \\ \frac{\partial v}{\partial t} &= d_v \Delta v + g(u, v) \, , \end{aligned} \tag{A3.3.1}$$

where $d_u, d_v > 0$ are the respective diffusion coefficients, $f, g \in C^1$ are the reaction terms, and the Laplace delta is $\Delta = \partial^2/\partial x^2 + \partial^2/\partial y^2$. No flux boundary conditions are attached; denoting the directional derivative in the direction orthogonal to the boundary ∂B by $\partial/\partial n$:

$$\frac{\partial u(t, x, y)}{\partial n} = \frac{\partial v(t, x, y)}{\partial n} = 0 \, , \; t \geq 0 \, , \; (x, y) \in \partial B \, . \tag{A3.3.2}$$

The ODE system

$$\dot{u} = f(u, v) \, , \; \dot{v} = g(u, v) \, , \tag{A3.3.3}$$

where the overdot denotes differentiation with respect to time t, is called the *kinetic system* attached to Eq. (A3.3.1). If $f(\bar{u}, \bar{v}) = g(\bar{u}, \bar{v}) = 0$ then the equilibrium $(\bar{u}, \bar{v})$ is also a constant solution of the PDE system (A3.3.1). However, contrary to the scalar case the asymptotic stability of $(\bar{u}, \bar{v})$ with respect to system (A3.3.3) does not necessarily imply its asymptotic stability with respect to the PDE system (A3.3.1).

DEFINITION A3.3.1. We say that a stationary solution $(\tilde{u}(x,y), \tilde{v}(x,y))$ of the problem (A3.3.1)-(A3.3.2) is *stable in the Lyapunov sense* if for every $\varepsilon > 0$ there exists a $\delta > 0$ such that if for the initial value $(u(0,x,y), v(0,x,y))$ of a solution of the problem

$$\sup_B (|u(0,x,y) - \tilde{u}(x,y)| + |v(0,x,y) - \tilde{v}(x,y)|) < \delta$$

holds then for $t > 0$, $(x,y) \in B$ we have

$$|u(t,x,y) - \tilde{u}(x,y)| + |v(t,x,y) - \tilde{v}(x,y)| < \varepsilon ;$$

we say that this stationary solution is *asymptotically stable* if it is stable in the Lyapunov sense and there exists an $\eta > 0$ such that if

$$\sup_B (|u(0,x,y) - \tilde{u}(x,y)| + |v(0,x,y) - \tilde{v}(x,y)|) < \eta$$

then

$$\lim_{t \to \infty} (|u(t,x,y) - \tilde{u}(x,y)| + |v(t,x,y) - \tilde{v}(x,y)|) = 0 .$$

Even if B is convex for problem (A3.3.1)-(A3.3.2) the situation may arise that *a constant stationary solution is asymptotically stable with respect to the kinetic system (A3.3.3) but is unstable with respect to problem (A3.3.1)-(A3.3.2).* This possibility, discovered by Turing (1952) in his famous paper on morphogenesis is called *Turing (*or *diffusion driven) instability.* Spatially constant initial conditions are, at the same time, initial conditions with respect to the kinetic system, and the corresponding solution of the latter is obviously also a solution of the problem (A3.3.1)-(A3.3.2) depending only on the time t. Therefore, Turing instability means that solutions with constant initial values near to the constant solution tend to the latter as time tends to infinity while solutions corresponding to spatially nonconstant initial conditions arbitrarily near to the constant solution may tend away from it. This phenomenon is interesting because the general experience is that diffusion helps stability by evening out differences, and now the opposite happens, and it is also of interest because Turing instability may go together with the occurrence of a spatially nonconstant stationary solution, which is called a *pattern.*

First, necessary conditions will be deduced for the occurrence of Turing instability. Denote an equilibrium point of the kinetic system (A3.3.3) by $(\bar{u}, \bar{v})$, linearize the system at this point, denote the coefficient matrix of the linearized system by

$$A = \begin{bmatrix} f'_u(\bar{u},\bar{v}) & f'_v(\bar{u},\bar{v}) \\ g'_u(\bar{u},\bar{v}) & g'_u(\bar{u},\bar{v}) \end{bmatrix} ,$$

and assume that the eigenvalues of this matrix μ_1^0, μ_2^0 have negative real parts. This assumption implies the asymptotic stability of the equilibrium with respect to system (A3.3.3). The characteristic polynomial of A is

$$\mu^2 - (f'_u + g'_v)\mu + \det A \tag{A3.3.4}$$

so that from the remark following Theorem A1.1.2 our assumption is equivalent to saying that

$$f'_u + g'_v < 0\ ,\ \det A = f'_u g'_v - f'_v g'_u > 0 \tag{A3.3.5}$$

(here and in the sequel the arguments will not be written out; they are always $(\bar{u}, \bar{v})$). Now system (A3.3.1) is to be linearized at the constant stationary solution $(\bar{u}, \bar{v})$, the "general solution" of the linearized system with boundary conditions (A3.3.2) will be written out, and the stability of the $(0,0)$ solution of the linearized problem will be considered. By shifting the origin of the phase space into $(\bar{u}, \bar{v})$, introducing the coordinates $p = u - \bar{u}$, $q = v - \bar{v}$, writing out the equation in the new coordinates, and dropping the higher-order terms, we arrive at the linearized system

$$\begin{aligned} \frac{\partial p}{\partial t} &= d_u \Delta p + f'_u p + f'_v q \\ \frac{\partial q}{\partial t} &= d_v \Delta q + g'_u p + g'_v q \end{aligned} \tag{A3.3.6}$$

to which the boundary conditions

$$\frac{\partial p(t,x,y)}{\partial n} = \frac{\partial q(t,x,y)}{\partial n} = 0\ ,\ t \geq 0\ ,\ (x,y) \in \partial B \tag{A3.3.7}$$

are attached. The problem (A3.3.6)-(A3.3.7) is to be solved by Fourier's method. We suppose that the problem has a solution of the form $(T_1(t), T_2(t))R(x,y)$, substitute this into Eq. (A3.3.6), divide the first and the second equation by $T_1 R$ and $T_2 R$, respectively, and obtain

$$\begin{aligned} \frac{\dot{T}_1}{T_1} &= d_u \frac{\Delta R}{R} + f'_u + f'_v \frac{T_2}{T_1} \\ \frac{\dot{T}_2}{T_2} &= d_v \frac{\Delta R}{R} + g'_u \frac{T_1}{T_2} + g'_v \end{aligned}$$

or

$$\begin{aligned} \frac{1}{d_u}\left(\frac{\dot{T}_1}{T_1} - f'_u - f'_v \frac{T_2}{T_1}\right) &= \frac{\Delta R}{R} = -\lambda \\ \frac{1}{d_v}\left(\frac{\dot{T}_2}{T_2} - g'_u \frac{T_1}{T_2} - g'_v\right) &= \frac{\Delta R}{R} = -\lambda\ , \end{aligned} \tag{A3.3.8}$$

where λ is a constant because the first term depends only on t and the second only on x, y. First the boundary value problem

$$\Delta R = -\lambda R \ , \ \frac{\partial R(x,y)}{\partial n} = 0 \ , \ (x,y) \in \partial B \qquad \text{(A3.3.9)}$$

is to be solved. It is known (see Vladimirov, 1967; or Evans, 1998) that (A3.3.9) has a countable set of nonnegative eigenvalues $0 = \lambda_0 \leq \lambda_1 \leq \lambda_2 \leq ... \leq \lambda_k \leq ...$ with eigenfunctions $R_k(x,y)$, $(k = 0, 1, 2, ...)$ such that the eigenfunctions belonging to different eigenvalues are orthogonal to each other. To be sure, the determination of the eigenvalues and the eigenfunctions might be a difficult problem depending on the geometry of B but for simple domains such as a rectangle, a rectangular triangle or a circle, it can be done explicitly. Second, the 2D ODE system for (T_1, T_2) is written out from Eq. (A3.3.8):

$$\begin{aligned} \dot{T}_1 &= (f'_u - \lambda d_u)T_1 + f'_v T_2 \\ \dot{T}_2 &= g'_u T_1 + (g'_v - \lambda d_v)T_2 \end{aligned}$$

or

$$\begin{bmatrix} \dot{T}_1 \\ \dot{T}_2 \end{bmatrix} = (A - \lambda D) \begin{bmatrix} T_1 \\ T_2 \end{bmatrix}, \qquad \text{(A3.3.10)}$$

where D =diag$[d_u, d_v]$ is the diffusion matrix. We have to substitute each eigenvalue of problem (A3.3.9) into Eq. (A3.3.10) for λ and solve the latter system. The characteristic polynomial of (A3.3.10) is

$$\det(A - \lambda D - \mu I) = \mu^2 - \mu(f'_u + g'_v - \lambda(d_u + d_v))$$

$$+ \det A - \lambda(d_u g'_v + d_v f'_u) + \lambda^2 d_u d_v \ . \qquad \text{(A3.3.11)}$$

Using (A3.3.5) for $\lambda \geq 0$: Trace$(A - \lambda D) = f'_u + g'_v - \lambda(d_u + d_v)$ is negative and thus we may have instability only if

$$\det(A - \lambda D) = \det A - \lambda(d_u g'_v + d_v f'_u) + \lambda^2 d_u d_v \leq 0 \ . \qquad \text{(A3.3.12)}$$

If the characteristic polynomial has a double root then as the product of the roots this determinant is positive, so that in this case stability prevails. Therefore, we may assume without loss of generality that for each λ_k $(k = 0, 1, 2, ...)$ (A3.3.11) has two distinct roots μ_1^k, μ_2^k with eigenvectors s^{k1}, s^{k2}, respectively, and as a consequence, the general solution of (A.3.3.10) is

$$\begin{bmatrix} T_1(t) \\ T_2(t) \end{bmatrix} = c_{k1} e^{\mu_1^k t} s^{k1} + c_{k2} e^{\mu_2^k t} s^{k2} \ ,$$

where c_{k1}, c_{k2} are arbitrary constants. This way, for each $k = 0, 1, 2, ...$ we have obtained a solution of the boundary value problem (A3.3.6)-(A3.3.7):

$$R_k(x,y)(c_{k1} e^{\mu_1^k t} s^{k1} + c_{k2} e^{\mu_2^k t} s^{k2}) \ .$$

If the series

$$\sum_{k=0}^{\infty} R_k(x,y)(c_{k1}e^{\mu_1^k t}s^{k1} + c_{k2}e^{\mu_2^k t}s^{k2}) \tag{A3.3.13}$$

is convergent and can be substituted into Eq. (A3.3.6), then it is also a solution of this linear problem. If smooth initial conditions $(p(0,x,y), q(0,x,y)) = (P(x,y), Q(x,y))$ are given then, in order to determine the appropriate coefficients in the last series, the initial functions are to be expanded into generalized Fourier series:

$$\begin{bmatrix} P(x,y) \\ Q(x,y) \end{bmatrix} = \sum_{k=0}^{\infty} R_k(x,y)(c_{k1}s^{k1} + c_{k2}s^{k2}) \, .$$

With the coefficients determined this way we get the solution of the boundary-initial value problem:

$$\begin{bmatrix} p(t,x,y) \\ q(t,x,y) \end{bmatrix} = \sum_{k=0}^{\infty} R_k(x,y)(c_{k1}e^{\mu_1^k t}s^{k1} + c_{k2}e^{\mu_2^k t}s^{k2}) \, .$$

If for each $\lambda_k, (k = 0, 1, 2, ...)$ system (A3.3.10) is asymptotically stable, then in the last series every term tends to zero exponentially as time tends to infinity and hence every solution of problem (A3.3.6)-(A3.3.7) tends to (0,0). From Casten and Holland (1977) the asymptotic stability of the constant solution $(\bar{u}, \bar{v})$ of the nonlinear problem (A3.3.1)-(A3.3.2) is implied. In order to have Turing instability system (A3.3.10) must be unstable for at least one λ_k. This means that (A3.3.12) must hold for some positive λs. The roots of the polynomial in (A3.3.12) are

$$\lambda^{1,2} = \frac{1}{2d_u d_v}\left(d_u g'_v + d_v f'_u \pm \sqrt{(d_u g'_v + d_v f'_u)^2 - 4d_u d_v \det A}\right) \, .$$

If $d_u g'_v + d_v f'_u \leq 0$, then no root is positive. We have positive roots only if

$$d_u g'_v + d_v f'_u > 0 \tag{A3.3.14}$$

and

$$(d_u g'_v + d_v f'_u)^2 > 4d_u d_v \det A \, . \tag{A3.3.15}$$

In this case both roots are positive and distinct (a double positive root is of no use because then we do not have an interval where (A3.3.12) holds). This way we arrived at

THEOREM A3.3.1 *If the equilibrium solution $(\bar{u}, \bar{v})$ is Turing unstable then conditions (A3.3.5), (A3.3.14), and (A3.3.15) must hold.*

By the first condition of Eq. (A3.3.5) at least one of f'_u, g'_v must be negative. By (A3.3.14) at least one of them must be positive. Hence, $f'_u g'_v < 0$. We may assume without loss of generality that

$$f'_u > 0, \; g'_v < 0 \; . \tag{A3.3.16}$$

This means that u acts as an activator (its increase is increasing its production) and v as an inhibitor (its increase is decreasing its production). Then by (A3.3.5) $0 < f'_u < -g'_v$ and by (A3.3.14) $f'_u > -g'_v d_u/d_v$. The last two inequalities imply that *Turing instability may occur only if* $0 < d_u/d_v < 1$, that is, *the diffusion coefficient of the inhibitor is larger than that of the activator*:

$$d_u < d_v \; . \tag{A3.3.17}$$

In the discussion that follows we assume that (A3.3.5) and (A3.3.14)-(A3.3.17) hold. Under these conditions $0 < \lambda^1 < \lambda^2$, so that if for some $k \geq 1$ the corresponding eigenvalue λ_k of Eq. (A3.3.9) falls into this interval, then (A3.3.12) holds, that is, $\mu_1^k \mu_2^k \leq 0$, which implies that, for example, $\mu_1^k < 0 \leq \mu_2^k$. As a consequence, the corresponding term in series (A3.3.13) does not tend to zero, that is, the constant solution is no longer asymptotically zero. If μ_2^k is positive then this term tends to infinity; if $\mu_2^k = 0$ then $R_k(x,y)s^{k2}$ is a spatially nonconstant stationary solution, that is, a *pattern*.

Suppose that one of the parameters in system (A3.3.1) is varied in such a way that this does not affect conditions (A3.3.5), that is, the equilibrium point $(\bar{u}, \bar{v})$ remains an asymptotically stable solution of the kinetic system (A3.3.3). This *bifurcation parameter* denoted in the sequel by b can be one of the diffusion coefficients, a measure of the spatial domain B (its diameter or area), or a parameter in the functions f and g. As the bifurcation parameter is varied the interval $[\lambda^1, \lambda^2] \subset \mathbb{R}_+$ sweeps through the positive axis λ, changing its length and in the process maybe engulfing some of the eigenvalues λ_k. *Suppose that for $b < b_0$ the interval $[\lambda^1, \lambda^2]$ does not contain any of the λ_k-s: $[\lambda^1, \lambda^2] \cap \{\lambda_1, \lambda_2, \lambda_3, \ldots\} = \emptyset$,the empty set, at $b = b_0$ we have $\lambda_i = \lambda^2$, say, and for $b > b_0$ the interval contains this eigenvalue: $\lambda_i \in (\lambda^1, \lambda^2)$; we say then that at b_0 the constant solution $(\bar{u}, \bar{v})$ undergoes a Turing bifurcation.* This means that for $b < b_0$ the constant solution is asymptotically stable with respect to problem (A3.3.1)-(A3.3.2), for $b > b_0$ it is unstable while it remains asymptotically stable with respect to the kinetic system, and at $b = b_0$ the linearized problem (A3.3.6)-(A3.3.7) has a spatially nonconstant stationary solution. Applying a theorem from Smoller (1983), under generic conditions one may prove that in this case *in a (possibly one-sided) neighborhood of b_0 the nonlinear problem (A3.3.1)-(A3.3.2) has a spatially nonconstant stationary solution, a pattern* (for a proof in case of a concrete model see Cavani and Farkas 1994).

Appendix 4

RIEMANNIAN GEOMETRY

A4.1 Local Riemannian Geometry

Riemannian geometry was introduced by Riemann in 1854 as the n-dimensional generalization of the theory of curved surfaces of the 3D Euclidean space. For more than 50 years it was more or less a "Sleeping Beauty" until it turned out that it was indispensable in Einstein's general theory of relativity because in 4D space-time the metric is not Euclidean but Riemannian and varies from point to point due to the distribution of masses in the Universe. In modern treatments of Riemannian geometry there is an underlying abstract topological space (sometimes infinite dimensional) that is the countable union of sets (called coordinate charts) diffeomorphic to $\mathbb{R}^n$ (to a Hilbert space in the infinite dimensional case). We shall not need that here. All we need is a set V (an abstract topological space) that can be mapped in a one-to-one way to a single copy of $\mathbb{R}^n$ (or to an open connected subset of $\mathbb{R}^n$). We need the set V in order to be able to speak about points (the elements of V) independently of the coordinate system, the latter being introduced by the mapping of V to $\mathbb{R}^n$. For a more general treatment see, for example, Klingenberg (1978).

Thus, if a point $\mathbf{x} \in V$ is given we represent it by its image $x = (x^1, x^2, ..., x^n)$ in $\mathbb{R}^n$ (putting the indices up simplifies the calculations here). In this Appendix *smooth* means three times continuously differentiable. We allow smooth coordinate transformations, that is, C^3 diffeomorphisms $\phi : \mathbb{R}^n \mapsto \mathbb{R}^n$ denoting the image of $x \in \mathbb{R}^n$ by $\tilde{x} = \phi(x) = (\tilde{x}^1, \tilde{x}^2, ..., \tilde{x}^n) \in \mathbb{R}^n$. In the new coordinate system $\mathbf{x} \in V$ is represented by $\tilde{x}$. To each point $\mathbf{x}$ of V we make to correspond an n-dimensional linear space T_xV and call it the *tangent space at* $\mathbf{x}$. The coordinate system x induces a basis, the *natural basis* in T_xV, consisting of the base vectors $\mathbf{u}_1$ =col$[1, 0, ..., 0]$, $\mathbf{u}_2$ =col$[0, 1, ..., 0]$, ..., $\mathbf{u}_n$ =col$[0, 0, ..., 1]$. Intuitively these are the tangent vectors of the coordinate lines $x^1, x^2, ..., x^n$, respectively, that is, the curves along which only the respective coordinate varies and the other $n-1$ coordinates are constant. A vector $\mathbf{v}$ of the space T_xV is called a vector of the space V at point $\mathbf{x}$. In a natural basis its coordinates are v =col$[v^1, v^2, ..., v^n]$, that is, $\mathbf{v} = \sum_{k=1}^n v^k \mathbf{u}_k$. (In this Section all summations will go from 1 to n, so the limits of the summation will not be written out in the sequel; if necessary the *index* for which the summation goes will be denoted.) In the new coordinate system $\tilde{x}$ the coordinates of the vector $\mathbf{u}_k$ of the natural basis induced by the old coordinate system x are $[\partial\tilde{x}^1/\partial x^k, ..., \partial\tilde{x}^n/\partial x^k]$; in other words this means that $\mathbf{u}_k = \sum_i \tilde{\mathbf{u}}_i \partial\tilde{x}^i/\partial x^k$, where $\tilde{\mathbf{u}}_i$ are the vectors of the natural basis of the new coordinate system. Substituting this into the expression of the vector $\mathbf{v}$ gives us $\mathbf{v} = \sum_k v^k \sum_i \tilde{\mathbf{u}}_i \partial\tilde{x}^i/\partial x^k = \sum_{i,k} \tilde{\mathbf{u}}_i v^k \partial\tilde{x}^i/\partial x^k$. On the other hand, in the new system the same vector has the representation $\mathbf{v} = \sum_i \tilde{v}^i \tilde{\mathbf{u}}_i$. Hence, we see that the coordinates of a vector are transformed by the formula

$$\tilde{v}^i = \sum_k v^k \partial\tilde{x}^i/\partial x^k , \tag{A4.1.1}$$

or if the notations $\partial\tilde{x}/\partial x = \left[\partial\tilde{x}^i/\partial x^k\right]$ are introduced for the matrix of the coordinate transformation in T_xV and $\tilde{v}$ =col$[\tilde{v}^1, \tilde{v}^2, ..., \tilde{v}^n]$:

$$\tilde{v} = (\partial\tilde{x}/\partial x)v . \tag{A4.1.1$'$}$$

If $x(t) = (x^1(t), ..., x^n(t))$ is a smooth curve in V (t varying in an interval) then at the point $x(t) \in V$ the derivative $dx(t)/dt$ =col$[\dot{x}^1(t), ..., \dot{x}^n(t)]$ is a vector in $T_{x(t)}V$ called the *tangent vector of the curve at the point $x(t)$*. In the new coordinate system the equation of the curve is given by $\tilde{x}(t) = \phi(x(t))$, and its tangent vector is $d\tilde{x}(t)/dt$ =col$[\tilde{x}^{1\cdot}(t), ..., \tilde{x}^{n\cdot}(t)] = (\partial\tilde{x}/\partial x)dx(t)/dt$ by the chain rule and in conformity with Eq. (A4.1.1$'$).

The function $a : T_xV \times T_xV \mapsto \mathbb{R}$ is called a *tensor* if it is bilinear, that is, if for arbitrary $\mathbf{u}, \mathbf{v}, \mathbf{w} \in T_xV$ and α, β we have

$$\begin{aligned} a(\alpha\mathbf{u}+\beta\mathbf{v}, \mathbf{w}) &= \alpha a(\mathbf{u}, \mathbf{w})+\beta a(\mathbf{v}, \mathbf{w}) , \\ a(\mathbf{w},\alpha\mathbf{u}+\beta\mathbf{v}) &= \alpha a(\mathbf{w}, \mathbf{u})+\beta a(\mathbf{w}, \mathbf{v}) . \end{aligned}$$

We say that a tensor g is *symmetric* if for every $\mathbf{u}, \mathbf{v} \in V$ there holds $g(\mathbf{u}, \mathbf{v}) = g(\mathbf{v}, \mathbf{u})$. We say that a symmetric tensor g is *positive definite* if for every $\mathbf{v} \in V, \mathbf{v} \neq \mathbf{0}$ there holds $g(\mathbf{v}, \mathbf{v}) > 0$. If a coordinate system x is given, inducing the natural basis in T_xV, then tensor a is given by its coordinate matrix $A = [a_{ik}]$ and for arbitrary $\mathbf{v}, \mathbf{w} \in T_xV$ we have $a(\mathbf{v}, \mathbf{w}) = \sum_{i,k} a_{ik} v^i w^k$ where v^i and w^k are the coordinates of the respective vectors. If the new coordinate system $\tilde{x}$ is introduced then the same value is given by $a(\mathbf{v}, \mathbf{w}) = \sum_{j,h} \tilde{a}_{jh} \tilde{v}^j \tilde{w}^h$ where $\tilde{A} = [\tilde{a}_{jh}]$ is the coordinate matrix of the tensor in the new system of coordinates. Substituting from the transformation formula (A4.1.1) of vector coordinates we have

$$\sum_{i,k} a_{ik} v^i w^k = a(\mathbf{v}, \mathbf{w}) = \sum_{j,h,i,k} \tilde{a}_{jh} \frac{\partial\tilde{x}^j}{\partial x^i} v^i \frac{\partial\tilde{x}^h}{\partial x^k} v^k$$

from where

$$a_{ik} = \sum_{j,h} \tilde{a}_{jh} \frac{\partial \tilde{x}^j}{\partial x^i} \frac{\partial \tilde{x}^h}{\partial x^k}$$

follows. Multiplying by $\partial x^i/\partial \tilde{x}^p$ and by $\partial x^k/\partial \tilde{x}^q$, summing for i, k, and taking into account that $\partial x/\partial \tilde{x} = [\partial x^i/\partial \tilde{x}^p]$ is the inverse matrix of $\partial \tilde{x}/\partial x$, we obtain the transformation formula for the tensor coordinates:

$$\tilde{a}_{pq} = \sum_{i,k} \frac{\partial x^i}{\partial \tilde{x}^p} \frac{\partial x^k}{\partial \tilde{x}^q} a_{ik} . \tag{A4.1.2}$$

If to each point $\mathbf{x} \in V$ a vector $\mathbf{v}(\mathbf{x}) \in T_x V$ is made to correspond and in any admissible coordinate system the function $v(x)$ is smooth we say that $\mathbf{v}(\mathbf{x})$ is a *smooth vector field* over V. Similarly, we speak about *smooth tensor fields* over V. Smooth vector and tensor fields also can be defined, naturally, over subsets of V.

DEFINITION A4.1.1. If a smooth, symmetric, positive definite tensor field $g(\mathbf{x}; \cdot, \cdot)$ is given over V and the *arc length* of a smooth curve $\mathbf{x}(t)$, $t \in [\alpha, \beta]$ is defined by

$$s = \int_\alpha^\beta \sqrt{\sum_{i,k} g_{ik}(x(t)) \dot{x}^i(t) \dot{x}^k(t)}dt \tag{A4.1.3}$$

in the coordinate system x, then we say that a *Riemannian metric* has been introduced in V or that V is a *Riemann space* with *metric tensor* g.

The Riemannian metric induces a metric in each tangent space $T_x V$: the *scalar product* of the vectors $\mathbf{v}, \mathbf{w} \in T_x V$ is defined by $\mathbf{v} \cdot \mathbf{w} = \sum_{i,k} g_{ik}(x) v^i w^k$ and, as a consequence, the *length of a vector* $\mathbf{v} \in T_x V$ is $|\mathbf{v}| = \sqrt{\sum_{i,k} g_{ik}(x) v^i v^k}$. The cosine of the *angle* α of the two nonzero vectors $\mathbf{v}$ and $\mathbf{w}$ is, by definition, $\cos\alpha := \mathbf{v} \cdot \mathbf{w} / (|\mathbf{v}|\,|\mathbf{w}|)$, from where $\mathbf{v} \cdot \mathbf{w} = |\mathbf{v}|\,|\mathbf{w}| \cos\alpha$.

The arc length, the length of a vector, and the scalar product of vectors at a point of the space are invariant with respect to coordinate transformation. Indeed, if the coordinates of the metric tensor g and the vectors $\mathbf{v}$ and $\mathbf{w}$, respectively, are denoted by $\tilde{g}_{jh}$, $\tilde{v}^j$ and $\tilde{w}^h$, respectively, in the new coordinate system $\tilde{x}$ then by the transformation laws (A4.1.1)-(A4.1.2)

$$\begin{aligned}
\mathbf{v} \cdot \mathbf{w} &= \sum_{j,h} \tilde{g}_{jh}(\tilde{x}) \tilde{v}^j \tilde{w}^h \\
&= \sum_{j,h} \sum_{i,k} \frac{\partial x^i}{\partial \tilde{x}^j} \frac{\partial x^k}{\partial \tilde{x}^h} g_{ik}(x) \sum_p v^p \frac{\partial \tilde{x}^j}{\partial x^p} \sum_q w^q \frac{\partial \tilde{x}^h}{\partial x^q} \\
&= \sum_{i,k,p,q} \left(\sum_{j,h} \frac{\partial x^i}{\partial \tilde{x}^j} \frac{\partial x^k}{\partial \tilde{x}^h} \frac{\partial \tilde{x}^j}{\partial x^p} \frac{\partial \tilde{x}^h}{\partial x^q} \right) g_{ik}(x) v^p w^q \\
&= \sum_{i,k,p,q} \delta_p^i \delta_q^k g_{ik} v^p w^q = \sum_{i,k} g_{ik}(x) v^i w^k
\end{aligned}$$

where δ^i_p is the *Kronecker delta:*

$$\delta^i_p = \begin{cases} 1, \text{if } i = p\,, \\ 0, \text{if } i \neq p\,. \end{cases}$$

Given vector $\mathbf{v}$ with its coordinates $[v^1, ..., v^n]$ at the point x of the Riemannian space V the quantities $v_i = \sum_k g_{ik}(x)v^k$, $(i = 1, 2, ..., n)$ are called the *covariant coordinates of vector* $\mathbf{v}$. Distinctively, the original coordinates are called the *contravariant coordinates of* $\mathbf{v}$. (The names are related to the transformation laws of the two kinds of coordinates.) Denoting the inverse matrix of $[g_{ik}]$ by $[g^{jh}]$, multiplying the previous equation by $g^{ji}(x)$, and summing for i we obtain the formula showing how one can regain the original coordinates of a vector from the covariant coordinates:

$$\sum_i g^{ji}(x)v_i = \sum_{i,k} g^{ji}(x)g_{ik}(x)v^k = \sum_k \delta^j_k v^k = v^j\,. \tag{A4.1.4}$$

Denoting the covariant coordinates of the vector in the coordinate system $\tilde{x}$ by $\tilde{v}_j$, the transformation formulae for these are obtained the following way by applying Eqs. (A4.1.1)-(A4.1.2):

$$\begin{aligned} \tilde{v}_j &= \sum_k \tilde{g}_{jk}(\tilde{x})\tilde{v}^k = \sum_k \sum_{i,h} \frac{\partial x^i}{\partial \tilde{x}^j}\frac{\partial x^h}{\partial \tilde{x}^k} g_{ih}(x) \sum_p v^p \frac{\partial \tilde{x}^k}{\partial x^p} \\ &= \sum_i \frac{\partial x^i}{\partial \tilde{x}^j} \sum_h g_{ih}(x)v^h = \sum_i \frac{\partial x^i}{\partial \tilde{x}^j} v_i\,. \end{aligned} \tag{A4.1.5}$$

Obviously, if at the point $x \in V$ the vector $\mathbf{v}$ is given by its covariant coordinates v_i and the vector $\mathbf{w}$ by its contravariant coordinates w^i, then the scalar product of the two vectors is $\mathbf{v} \cdot \mathbf{w} = \sum_i v_i w^i$.

Let $f : V \mapsto \mathbb{R}$ be a *smooth scalar field* over V. The level surface $S_c := \{\mathbf{x} \in V \mid f(\mathbf{x}) = c\}$ where c is a constant from the range of f is a *hypersurface* of space V. Let $\mathbf{x}(t)$, $t \in (\alpha, \beta)$ be a smooth curve in S_c; this means that $f(\mathbf{x}(t)) \equiv c$. In the coordinate system x we have $f(x(t)) \equiv c$. Differentiating the identity, we get $\sum_i \dot{x}^i \partial f/\partial x^i \equiv 0$. If the quantities $\nabla_i f := \partial f/\partial x^i$, $(i = 1, 2, ..., n)$ are considered as the covariant coordinates of a vector denoted by ∇f the last identity means that this vector is orthogonal to the tangent vector of the curve (of *any* curve) of the hypersurface S_c at the point $\mathbf{x}(t)$, because the scalar product $\dot{\mathbf{x}}(t) \cdot \nabla f(\mathbf{x}(t)) = 0$. To consider the quantities $\nabla_i f$ as covariant vector coordinates is justified by their transformation law when a new coordinate system $\tilde{x}$ is introduced. By the chain rule in the new system the coordinates are

$$\tilde{\nabla}_j f = \sum_i \frac{\partial f}{\partial x^i}\frac{\partial x^i}{\partial \tilde{x}^j}$$

in accordance with Eq. (A4.1.5). The smooth vector field ∇f is called the *gradient* of the scalar field f. As we have seen at every point it is orthogonal to the level surface of f passing through that point. The contravariant coordinates of the gradient are, by Eq. (A4.1.4),

$$\nabla f(x) = \left[\nabla^1 f(x), ..., \nabla^n f(x)\right] \;,$$
$$\nabla^i f(x) := \sum\nolimits_k g^{ik}(x) \nabla_k f(x) \;,\; (i = 1, 2, ...n) \,. \qquad \text{(A4.1.6)}$$

The *directional derivative* $\partial f / \partial u$ of the scalar field f at a point $\mathbf{x}$ in the direction $\mathbf{u} \in T_x V$ (where $\mathbf{u}$ is a unit vector: $g(\mathbf{x}; \mathbf{u}, \mathbf{u}) = 1$) is $\partial f / \partial u = \nabla f(\mathbf{x}) \cdot \mathbf{u}$. This is the rate of change of f along a curve passing through the point $\mathbf{x}$ in the given direction if the curve is parametrized by the arc length, that is, if its unit tangent vector at $\mathbf{x}$ is $\mathbf{u}$. It is clear that, the rate of change is the fastest in the direction of the gradient

$$\max_{|u|=1} \nabla f(\mathbf{x}) \cdot \mathbf{u} = |\nabla f(\mathbf{x})| \cos 0 = |\nabla f(\mathbf{x})| \;.$$

This is roughly what we need from Riemannian geometry. We note that algebraic manipulations between vectors belonging to tangent spaces of different points are impossible. This makes differentiation of vector and tensor fields difficult. However, the parallel displacement of vectors along curves can be defined and this makes differential calculus possible but this parallel displacement depends on the curve along which it is performed. It may happen, and in fact, this is the generic situation–that a vector is displaced parallelly along a closed curve and by returning to the starting point we do not get back the original vector. This phenomenon is related to the "curvature" of the space. The "curvature" of the Euclidean space is zero and as a consequence the parallel displacement there is absolute. In a Riemannian space those curves whose tangent vector is parallelly displaced along the curve itself are called *geodesics*. These curves play the role of the straight lines of an Euclidean space. Because the distance of two points of a Riemannian space is not directly defined we may look for the curve among the curves joining the two points whose arc length is minimal. It turns out that this curve is a geodesic. Its arc length is the *geodesic distance* of the two points.

If a metric tensor has been defined in the space by its coordinate matrix $[g_{ik}(x)]$ in a given coordinate system it may happen that an admissible coordinate transformation exists such that in the new coordinate system $\tilde{x}$ the matrix of the metric tensor is the unit matrix at every point:

$$\tilde{g}_{pq}(\tilde{x}) = \sum\nolimits_{i,k} \frac{\partial x^i}{\partial \tilde{x}^p} \frac{\partial x^k}{\partial \tilde{x}^q} g_{ik}(x(\tilde{x})) \equiv \delta_{pq} = \begin{cases} 1, & p = q \\ 0, & p \neq q \end{cases} \;.$$

If this is the case, then the scalar product of two vectors $\mathbf{v}, \mathbf{w}$ is $\mathbf{v} \cdot \mathbf{w} = \sum_i \tilde{v}^i \tilde{w}^i$ and the length of a vector is obtained the Pythagorean way: $|\mathbf{v}| = \left(\sum_i (\tilde{v}^i)^2\right)^{1/2}$; this means that the space is, in fact, a Euclidean space in which $\tilde{x}$ is a Cartesian orthogonal coordinate system while x was a "curvilinear coordinate system." Thus, the Euclidean space is a special Riemannian space. To be sure, in a "real Riemannian space" no such coordinate system can be found.

A4.2 Shahshahani's Metric

Consider the interior of the positive orthant of $\mathbb{R}^n$ denoted by $V = \text{Int}\,\mathbb{R}^n_+ = \{x \in \mathbb{R}^n \mid x^i > 0, i = 1, 2, ..., n\}$ and introduce the metric tensor

$$g_{ik}(x) := \delta_{ik} 1/x^i \ (i, k = 1, 2, ..., n) \tag{A4.2.1}$$

where δ_{ik} is the Kronecker delta. The metric introduced by this tensor is called *Shahshahani's metric* (see Shahshahani, 1979). The arc length of the smooth curve $x(t)$, $t \in [\alpha, \beta]$ as a function of the parameter is given in this space by

$$s(t) = \int_\alpha^t \sqrt{\sum_i \frac{1}{x^i(\tau)} (\dot{x}^i(\tau))^2 d\tau} \ .$$

At the points of the curve the differential of the arc length is

$$ds = \sqrt{\sum_i \frac{1}{x^i(t)} (\dot{x}^i(t))^2 dt} = \sqrt{\sum_i \frac{1}{x^i} (dx^i)^2}, \tag{A4.2.2}$$

where $(dx^i)^2$ is the square of the differential of the ith coordinate function of the curve. The approximate distance of two "neighboring" points x and $x + dx$ is given by Eq. (A4.2.2). This metric seems to be suitable if x is the genome type distribution in a population. The change dx^i in the quantity of genome type i is weighted by the square root of the reciprocal of the quantity x^i : if this genome type is in abundance then a small change in its quantity has no significance whatsoever but if it is scarce then a small change may drive it to or save it from extinction. Usually we are not interested in the actual numbers of the different genome types but in their respective *frequencies* in the population, that is, in those values of x that are in the simplex

$$S = \{x \in V \mid \sum_i x^i = 1, \ x^i > 0, \ i = 1, 2, ..., n\} \ .$$

The *average fitness* of the population (see Section 4.2) is

$$\bar{m}(x) = \sum_{i,k} m_{ik} x^i x^k \ , \ x \in S \ ,$$

where the m_{ik} are positive constants. We are to determine the gradient vector of the scalar field $\bar{m}$ at points of S and its projection to the tangent space of S. These calculations become considerably simpler if we realize that V equipped with Shahshahani's metric is, in fact, a Euclidean space. Indeed, the coordinate transformation

$$x^i = (\tilde{x}^i)^2/4 \ , \ \tilde{x}^i = 2\sqrt{x^i}, \ \partial x^i / \partial \tilde{x}^j = \delta^i_j \tilde{x}^i / 2 \tag{A4.2.3}$$

transforms the metric tensor into

$$\tilde{g}_{jh}(\tilde{x}) = \sum_i \delta^i_j \frac{\tilde{x}^i}{2} \delta^i_h \frac{\tilde{x}^i}{2} g_{ii}(x(\tilde{x})) = \delta_{jh} \frac{(\tilde{x}^j)^2}{4} \frac{1}{x^j(\tilde{x})} = \delta_{jh} \frac{(\tilde{x}^j)^2}{4} \frac{4}{(\tilde{x}^j)^2} = \delta_{jh} \ .$$

Thus, $\tilde{x}$ is a Cartesian orthogonal coordinate system in the space V while the original system x of the quantities is not. In the new system the average fitness is

$$\bar{m}(\tilde{x}) := \bar{m}(x(\tilde{x})) = \sum_{i,k} m_{ik} \left((\tilde{x}^i)^2/4\right)\left((\tilde{x}^k)^2/4\right) = \frac{1}{16}\sum_{i,k} m_{ik}(\tilde{x}^i)^2(\tilde{x}^k)^2 .$$

The equation of S in the Cartesian system is $\sum_i (\tilde{x}^i)^2 = 4$, $\tilde{x}^i > 0$, which means that in this metric S is a part of a sphere of radius 2. As seen from Eq. (A4.1.6) in a Euclidean space in a Cartesian system where $\tilde{g}^{ik} = \tilde{g}_{ik} = \delta_{ik}$, the covariant and contravariant coordinates of a vector are equal. Thus, the contravariant coordinates of the gradient of $\bar{m}$ are:

$$\tilde{\nabla}\bar{m}(\tilde{x}) = \frac{1}{4}\left[\sum_j m_{1j}\tilde{x}^1(\tilde{x}^j)^2, \sum_j m_{2j}\tilde{x}^2(\tilde{x}^j)^2, ..., \sum_j m_{nj}\tilde{x}^n(\tilde{x}^j)^2\right] .$$

The normal vector of the sphere S at the point $\tilde{x}$ is clearly $\tilde{x}$. Because the radius of the sphere is 2, this is the length of this vector and thus the normal unit vector of S at $\tilde{x}$ is $\tilde{x}/2$. Hence, the projection of the gradient to the tangent space $T_{\tilde{x}}S$ is

$$\begin{aligned}
\tilde{v}(\tilde{x}) : &= \tilde{\nabla}\bar{m}(\tilde{x}) - \left(\tilde{\nabla}\bar{m}(\tilde{x}) \cdot \frac{\tilde{x}}{2}\right)\frac{\tilde{x}}{2} \\
&= \tilde{\nabla}\bar{m}(\tilde{x}) - \bar{m}(\tilde{x})\tilde{x} \\
&= \left[\tilde{x}^1\left(\sum_j m_{1j}(\tilde{x}^j)^2/4 - \bar{m}(\tilde{x})\right), ..., \tilde{x}^n\left(\sum_j m_{nj}(\tilde{x}^j)^2/4 - \bar{m}(\tilde{x})\right)\right] .
\end{aligned}$$

Applying the tranformation formulae (A4.1.1) and (A4.2.3), the coordinates of this vector in the original coordinate system x are

$$\begin{aligned}
v^i(x) &= \frac{\partial x^i}{\partial \tilde{x}^i}\tilde{v}^i = \tilde{x}^i/2\left(\tilde{x}^i\left(\sum_j m_{ij}(\tilde{x}^j)^2/4 - \bar{m}(\tilde{x})\right)\right) \\
&= 2x^i\left(\sum_j m_{ij}x^j - \bar{m}(x)\right) \qquad (\text{ for } \sum_i x_i = 1) .
\end{aligned}$$

Compare with the right-hand side of Eq. (4.2.5).

References

Akin, E. (1979). *The Geometry of Population Genetics*, Lecture Notes in Biomaths 31, Berlin: Springer-Verlag.

Arnold, V.I. (1975). Critical points of smooth functions and their normal forms. *Uspehi. Mat. Nauk* **30**: 3-65 (Russian).

Beddington, J.R., Free, C.A. and Lawton, J.H. (1975). Dynamic complexity in predator-prey models framed in difference equations. *Nature* **255**: 58-60.

Capasso, V. (1993). *Mathematical Structures of Epidemic Systems*, Lecture Notes in Biomaths 97, Berlin: Springer-Verlag.

Casten, R.G. and Holland, C.J. (1977). Stability properties of solutions to systems of reaction-diffusion equations. *SIAM J. Appl. Math.* **33**: 353-364.

Casten, R.G. and Holland, C.J. (1978). Instability results for reaction-diffusion equations with Neumann boundary conditions. *J. Diff. Eq.* **27**: 266-273.

Cavani, M. and Farkas, M. (1994). Bifurcations in a predator-prey model with memory and diffusion II: Turing bifurcation. *Acta Math. Hungar.* **63**: 375-393.

Chow, S.N. and Hale, J.K. (1982). *Methods of Bifurcation Theory,* New York: Springer-Verlag.

Cooke, J. and Zeeman, E.C. (1976). A clock and wavefront model for control of the number of repeated structures during animal morphogenesis. *J. Theor. Biol.* **58**: 455-476.

Crow, J.F. and Kimura, M. (1970). *An Introduction to Population Genetics Theory,* New York: Harper and Row.

Cushing, J.M. (1977). *Integrodifferential Equations and Delay Models in Population Dynamics*, Lecture Notes in Biomaths, 20, Berlin: Springer-Verlag.

Czárán, T. (1998). *Spatiotemporal Models of Population and Community Dynamics*, London: Chapman and Hall.

DeLisi, Ch. (1983). Mathematical Modeling in Immunology. *Ann. Rev. Biophys. Bioeng.* **12**: 117-138.

Devaney, R.M. (1989). *An Introduction to Chaotic Dynamical Systems*, Redwood City: Addison-Wesley.

Dietz, K. and Hadeler, K.P. (1988). Epidemiological models for sexually transmitted diseases, *J. Math. Biol.* **26**: 1-25.

Eigen, M. and Schuster, P. (1979). *The Hypercycle: A Principle of Natural Self-organization,* Berlin: Springer-Verlag.

Ermentrout, B. (1991). Stripes or spots? Nonlinear effects in bifurcation of reaction-diffusion equations on the square, *Proc. R. Soc. London A*, **434**: 413-417.

Evans, L.C. (1998). *Partial Differential Equations*, Providence, RI: AMS.

Fargue, D. (1973). Réductibilité des systémes héréditaires á des systémes dinamiques. *C.R. Acad. Sci. Paris B*, **277**: 471-473.

Farkas, M. (1984a). Stable oscillations in a predator-prey model with time lag, *J. Math. Anal. Appl.*, **102**: 175-188.

Farkas, M. (1984b). Zip bifurcation in a competition model. *Nonlinear Analysis TMA* **8**: 1295-1309.

Farkas, M. (1987). Competitive Exclusion by Zip Bifurcation, in *Dynamical Systems, IIASA Workshop, Sopron,* Lecture Notes in Econ. and Math. Systems 287, Berlin: Springer-Verlag, 165-178.

Farkas, M. and Farkas, A. (1988a). Stable Oscillations in a More Realistic Predator-Prey Model with Time Lag, in *Asymptotic Methods of Mathematical Physics*, Kiev: Naukova Dumka, 250-256.

Farkas, M., Farkas, A. and Szabo, G. (1988). Multiparameter bifurcation diagrams in predator prey models with time lag. *J.Math. Biol.* **26**: 93-103.

Farkas, M. (1988b). On the Local Stability of n Predator (Prey) One Prey (Predator) Systems, in *Qualitative Theory of Differential Equations, Szeged (Hungary)*, Amsterdam: North Holland.

Farkas, M. and Freedman, H.I. (1989a). The stable coexistence of competing species on a renewable resource. *J. Math. Analysis Appl.* **138**: 461-472.

Farkas,M. and Freedman, H.I. (1989b). Stability conditions for two predator-one prey systems. *Acta Appl. Math.* **14**: 3-10.

Farkas, M. (1990). On the stability of one-predator two-prey systems. *Rocky Mountain J. Math.* **20**: 909-916.

Farkas, M. (1994). *Periodic Motions*, New York: Springer-Verlag.

Fisher, R.A. (1930). *The Genetical Theory of Natural Selection*, Oxford: Clarendon Press.

Freedman, H.I. (1976). Graphical stability, enrichment, and pest control by a natural enemy. *Math. Biosci.* **31**: 207-225.

Freedman, H.I. (1980). *Deterministic Mathematical Models in Population Ecology*, New York: Dekker.

Freedman, H.I. and Moson, P. (1990). Persistence definitions and their connections. *Proc. Amer. Math. Soc.* **109**: 1025-1033.

Guckenheimer, J. and Holmes, P.J. (1983). *Nonlinear Oscillations, Dynamical Systems and Bifurcations of Vector Fields,* New York: Springer-Verlag.

Hale, J.K. (1977). *Theory of Functional Differential Equations,* New York: Springer-Verlag.

Hartman, P. (1964). *Ordinary Differential Equations,* New York: Wiley.

Hethcote, H.W. and Yorke J.A. (1984). *Gonorrhea Transmission Dynamics and Control,* Lecture Notes in Biomaths 56, Berlin: Springer-Verlag.

Hofbauer, J. (1981). General cooperation theorem for hypercycles. *Monatsch. Math.* **91**: 233-240.

Hofbauer, J. and Sigmund, K. (1988). *The Theory of Evolution and Dynamical Systems,* Cambridge: Cambridge University Press.

Hofbauer, J. and Sigmund, K. (1998). *Evolutionary Games and Population Dynamics,* Cambridge: Cambridge University Press.

Hofbauer, J., Mallet-Paret, J. and Smith, H.L. (1991). Stable periodic solutions for the hypercycle system. *J. Dyn. Diff. Eq.* **3**: 423-436.

Hutson, V. and Moran, W. (1982). Persistence of species obeying difference equations. *J.Theoret. Biol.* **15**: 203-213.

Hutson, V. and Vickers, G.T. (1983). A criterion for permanent coexistence of species, with an application to a two-prey one-predator system. *Math. Biosci.* **63**: 253-269.

Jeffries, C., Klee, V. and van den Driessche, P. (1977). When is a matrix sign stable? *Can. J. Math.* **29**: 315-326.

Karlin, S. and Taylor, H.M. (1975). *A First Course in Stochastic Processes,* New York: Academic Press.

Kauffman, S.A. (1993). *The Origins of Order,* Oxford: Oxford University Press.

Kermack, W.O. and McKendrick, A.G. (1927). Contributions to the mathematical theory of epidemics, *Proc. Royal Soc.* A, **115**: 700-721, (1932) **138**: 55-83, (1933) **141**: 94-122.

Keyfitz, N. (1985). *Applied Mathematical Demography,* New York: Springer-Verlag.

Kirlinger, G. (1986). Permanence in Lotka-Volterra equations: linked prey-predator systems. *Math. Biosci.* **82**: 165-191.

Klingenberg, W. (1978). *A Course in Differential Geometry,* New York: Springer-Verlag.

Kocak, H. (1989). *Differential and Difference Equations through Computer Experiments,* New York: Springer-Verlag.

Langer, W.L. (1964). The Black Death, *Scientific American,* (February) 114-121.

Leslie, P.H. (1945). On the use of matrices in certain population mathematics. *Biometrika (London)* **33**: 183-212.

Leslie, P.H. (1948). Some further notes on the use of matrices in population mathematics. *Biometrika (London)* **35**: 213-245.

Lotka, A.J. (1924). *Elements of Mathematical Biology,* reprinted 1956, New York: Dover.

Lyons, M.J. and Harrison, L.G. (1991). A class of reaction-diffusion mechanisms which preferentially select striped patterns. *Chemical Physics Letters* **183**: 158-164.

Lyons, M.J. and Harrison, L.G. (1992). Stripe selection: An intrinsic property of some pattern-forming models with nonlinear dynamics. *Developmental Dynamics* **195**: 201-215.

Malthus, T.R. (1798). *An Essay on the Principle of Population,* reprinted 1970, Harmondsworth: Penguin.

Marchuk, G.I. (1983). *Mathematical Models in Immunology,* Ne York: Optimization Software.

Marchuk, G.I. (1997). *Mathematical Modeling of Immune Response in Infectious Diseases*, Maths. and its Applications, 395, Dordrecht: Kluwer.

Marcus, M. and Minc, H. (1964). *A Survey of Matrix Theory and Matrix Inequalities*, Boston: Allyn and Bacon.

May, R.M. and Oster G.F. (1976). Bifurcations and dynamic complexity in simple ecological models. *Amer. Natur.* **110**: 573-599.

May, R.M. (1981). *Theoretical Ecology*, Oxford: Blackwell.

Maynard Smith, J. and Szathmáry, E. (1995). *The Major Transitions in Evolution*, Oxford: Freeman.

McGehee, R. and Armstrong, R.A. (1977). Some mathematical problems concerning the ecological principle of competitive exclusion. *J. Diff. Eq.* **23**: 30-52.

Metz, J.A.J. and Diekmann, O. (1986). *The Dynamics of Physiologically Structured Populations*, Lecture Notes in Biomaths, 68, Berlin: Springer-Verlag.

Meinhardt, H. and Klinger, M. (1987). A model for pattern formation on the shells of molluscs. *J. Theoret. Biol.* **126**: 63-89.

Meinhardt, H. (1995). *The Algorithmic Beauty of Sea Shells*, Berlin: Springer-Verlag.

Murray, J.D. (1981). On pattern formation mechanisms for lepidopteran wing patterns and mammalian coat markings. *Phil. Trans. Roy. Soc. London B*, **295**: 473-496.

Murray, J.D. (1989). *Mathematical Biology*, Berlin: Springer-Verlag.

Okubo, A. (1980). *Diffusion and Ecological Problems: Mathematical Models*, Berlin: Springer-Verlag.

Peixoto, M.M. (1959). On structural stability. *Annals of Mathematics* **69**: 199-222.

Peixoto, M.M. (1962). Structural stability on two dimensional manifolds. *Topology* **1**: 101-120.

Perelson, A.S. (1989). Immune network theory. *Immunological Reviews* **110**: 5-36.

Poston, T. and Stewart, I. (1978). *Catastrophe Theory and its Applications*, London: Pitman.

Rosenzweig, M.L. and MacArthur, R.H. (1963). Graphical representation and stability conditions of predator-prey interaction. *Amer. Natur.* **47**: 209-223.

Rouche, N. and Mawhin, J. (1973). *Équations différentielles ordinaires I.II.*, Paris: Masson.

Rouche, N., Habets, P. and Laloy, M. (1977). *Stability Theory by Lyapunov's Direct Method*, New York: Springer-Verlag.

Rowe, G. (1994). *Theoretical Models in Biology*, Oxford: Clarendon Press.

Schuster, P., Sigmund, K. and Wolff, R. (1979). Dynamical systems under constant organization III. *J. Diff.Eq.* **32**: 357-368.

Shahshahani, S. (1979). A new mathematical framework for the study of linkage and selection. *Memoirs Amer. Math. Soc.* **211**.

Smale, S. (1961). On gradient dynamical systems. *Annals of Mathematics* **74**: 199-206.

Smith, Hal L. (1995). *Monotone Dynamical Systems, An Introduction to the Theory of Competitive and Cooperative Systems*, Providence: Amer. Math. Soc.

Smoller, J. (1983). *Shock Waves and Reaction-Diffusion Equations,* Berlin: Springer-Verlag.

Stépán, G. (1986). Great delay in a predator-prey model. *Nonlin. Anal. TMA* **10**: 913-929.

Svirezhev, Yu.M. and Logofet, D.O. (1983). *Stability of Biological Communities,* Moscow: Mir.

Szabo, G. (1987). A remark on M. Farkas: Stable oscilations in a predator-prey model with lag. *J. Math. Anal. Appl.* **128**: 205-206.

Takeuchi, Y. (1996). *Global Dynamical Properties of Lotka-Volterra Systems,* Singapore: World Scientific.

Thom, R. (1972). *Stabilité structurelle et morphogénèse,* Reading, MA: Benjamin.

Thomas, D. (1975). Artificial enzyme membranes, transport, memory and oscillatory phenomena, in *Analysis and Control of Immobilized Enzyme Systems,* Berlin: Springer-Verlag, 115-150.

Trotman, D. and Zeeman, E.C. (1976). The classification of elementary catastrophes of codimension ≤ 5, in *Lecture Notes in Maths. 525,* Berlin: Springer-Verlag, 263-327.

Turing, A.M. (1952). The chemical basis of morphogenesis. *Phil. Trans. Roy. Soc. London B,* **237**: 37-72 (reprinted: *Bull. Math. Biol.* **52**: 153-197, 1990).

Tusnády, G. (1995). Statistical analysis and prediction of Hungarian mortality curves, in *Human Resources and Social Stability during Transition in Hungary,* É. Erlich, and G. Révész, eds., San Francisco, CA: 242-253 Internat. Center of Economic Growth, 242-253.

Vladimirov, V.S. (1967). *The Equations of Mathematical Physics,* Moscow: Nauka.

Volterra, V. (1931). *Lecons sur la théorie mathématique de la lutte pour la vie,* Paris: Gauthier-Villars.

Watson, J.D. (1981). *The Double Helix: A Personal Account of the Discovery of the Structure of DNA,* London: Weidenfeld and Nicolson.

Willems, J.L. (1970). *Stability Theory of Dynamical Systems,* New York: Wiley.

Wolpert, L. (1981). Positional information and pattern formation. *Phil. Trans. Roy. Soc. London B* **295**: 441-450.

Zeeman, E.C. (1975). Catastrophe theory and biological patterns. *Ann. Rev. Biophys. Bioeng.* **4**: 210-215.

Index